Moral Acquaintances

Moral Acquaintances

METHODOLOGY IN BIOETHICS

Kevin Wm. Wildes, S. J.

University of Notre Dame Press
Notre Dame, Indiana

Manufactured in the United States of America

Library of Congress Cataloging-in-Publication Data

Wildes, Kevin Wm. (Kevin William), 1954–
Moral acquaintances : methodology in bioethics / Kevin William Wildes.
p.   cm.
Includes bibliographical references (p.) and index.
ISBN 0-268-03450-8 (cl : alk. paper) — ISBN 0-268-03452-4 (pa : alk. paper)
1. Medical ethics—Methodology. 2. Bioethics—Methodology. I. Title.
R725.5 .W55 2000
174′.2—dc21
00-036489

A faithful friend is a sturdy shelter:
the one who finds one finds a treasure.
A faithful friend is a life saving remedy
that no sum can measure.

Sirach 6:14

FOR MY MOTHER AND DAD

who have been faithful friends

CONTENTS

vii

PREFACE

One does not undertake a project like this book without a home. For me a home is a place of security and familiarity from which one can roam and explore the world. This book and I have had many wonderful homes that have made our roamings possible and I want to thank many people who have opened the doors of their homes to me during this project.

The first intellectual home for this book was provided during my graduate studies at Rice University and the Center for Ethics at Baylor College of Medicine. Baruch Brody and H. T. Engelhardt, Jr., patiently guided me in new intellectual explorations that shaped habits of thought and questioning that were essential for this project. I am forever grateful to them for what they did and for their friendship. Other colleagues, like Andy Lustig, Larry McCullough, and Gerry McKinney, provided daily conversations, questions, and encouragement. My friend Sarah Vaughan Brakman was a constant companion and source of encouragement to me.

The second intellectual home for this book has been at Georgetown University. At Georgetown I have had wonderful colleagues and students who have discussed, questioned, and encouraged me in this project. Tom L. Beauchamp has been very helpful as he has read, questioned, and discussed this manuscript with me. To him I am most grateful. Bob Veatch and Ed Pellegrino have read parts of the manuscript and I am grateful to them for their comments. Maggie Little and Madison Powers have been wonderful colleagues who have helped me articulate ideas and questions. LeRoy Walters, Director of the Kennedy Institute, and Wayne Davis, Chair of the Philosophy Department, have provided encouragement as well as research support and time to develop this project. I have had wonderful help in research from students like Rob Kennedy, Sarah Martin, and Mary Kay Scott. I have also had the wonderful support of the Kennedy Institute staff and the resources of our library and its ever gracious and expert staff.

In the development and writing of this book I have had lots of encouragement and support outside the academy. Monsignor Jim Jamail and the parish community of St. Vincent de Paul in Houston provided, literally, a home for me in the early stages of this project. The Jesuit Community of Georgetown University provided a home, and the encouragement, to move this project along to completion. A number of Jesuit friends have been very supportive of me. Jack Dennis, Tim Brown, Joe Sobierajski, John Swope, Ed Ingebretsen, Joe Ritzman, Brian Zinnamon, Leo O'Donovan, John Hollywood, Scott Pilarz, Steve Fields, Kevin Quinn, Jim Shea, and my departed friend John Ciani, have all helped me have a home while undertaking this project. Of course, I must thank my friends at Finley's Boxing Club in Washington, D.C., who helped to keep me sane and focused during this project!

Before I began this project I had learned the meaning of home from my family. My family has given me the security and love to explore the world with the knowledge that there was always a place to which I could return and just be myself. My dad and mother, Bill and Irene Wildes, gave me a home with security and love that allowed me to go out on this adventure I call a life. No words can express my debt to them.

In surveying all my friends and homes I remember the words of W. B. Yeats in his poem *The Municipal Gallery Revisited:* "Think of where man's glory most begins and ends, and say my glory was that I had such friends." For my friends, who have been home to me and this book, I have nothing but thanks.

Introduction:
Methodology and Bioethics

From decisions about the end of life to choices about the creation of life and, more recently, to questions concerning the cost and accessibility of health care, bioethics is a field of vigorous—and sometimes rancorous—public debate. Indeed the moral controversies and dilemmas of medicine and health care often propel bioethics into the headlines of newspapers and television talk shows. But the efforts of biomedical ethics to find resolutions and provide concrete guidance on these issues are not always so successful.

Bioethics is, then, a perplexing field. At times it seems to resolve difficult moral questions; at other times, it bogs down in endless dispute. The history of bioethics is filled with success. At its inception, the work of the National Commission for the Protection of Human Subjects of Biomedical and Behavioral Research led to guidelines and rules governing human research practices that gained widespread support. The commission's work has guided research and legislation for thirty years, and it has become a touchstone for daily medical practice.

By contrast the abortion "debate" in the United States appears to be a complete failure. Here no common ground has been achieved. The basic issues that were divisive in *Roe v. Wade* are still divisive, and the rancor of the controversy affects other questions in bioethics, most notably, fetal tissue research and preimplantation diagnosis. Bioethics as a field has not contributed greatly to the abortion issue, and other issues, such as physician-assisted suicide and the allocation of health care resources, touch off similar debates in the public forum.

Bioethics is also a relatively "new" field (see chapter 1), emerging only in the last four decades in response to the technological developments and research that define contemporary medicine and health care.[1] Among its subjects are physician-assisted suicide, organ transplantation and allocation, genetic research, and the distribution of health care resources. What, bioethics asks, is the moral choice among alternatives in these fields? What guidance can professional moralists, policy makers, and care givers offer to those who find themselves engaged in such issues?

Unlike most books in bioethics that offer guidance on one of these issues, for example, organ procurement or managed care, this book is not part of the standard repertory. The question it poses is not what we can do morally in a field of great moral controversy but how, in such cases, we determine a course of moral action. The book is, therefore, a book about methodology in a practice inspired by moral controversy and seminal experience.

In that sense, this book is entering new territory and inviting others to come along. Its focus, a careful examination of the different methods used in contemporary secular bioethics, is admittedly narrow, but its implications are potentially much wider. Dan Wickler has described the methodology of bioethics as completely anarchical.[2] And so I have found it. Bioethics is an interdisciplinary field that has many methods. Understanding the historical antecedents, conceptual commitments, and limitations underlying these methods can help us delineate the necessary conditions for bioethics in a secular society. From an exploration of different methods comes a better understanding of methodology. If, despite our best efforts, bioethics emerges from the exploration without having found a fundamental principle, its methodological chaos will at least be less chaotic.

Nor are we alone. Men and women bioethicists are beginning to ask these questions, as are others outside the field.[3] Still, it is fair to ask why. Why attempt to stir the murky depths of the methodological issue? Why not simply look at a particular issue such as physician-assisted suicide or patients' rights? Bioethics has not thought much about itself as a field or practice. One reason to do so now is that the emphasis on method can help us organize the discussion of particular issues. It can create a map of the controversies and how we might resolve them. If the exploration leads to engagement in real dialogue, it will also support a more sophisticated discussion. The move from particular moral controversies to methodology may help us predict the persistence of the controversy and suggest ways to reduce it.

Methodological exploration has another important consequence. As the field develops and becomes more "public" and professional, the social question arises about how we educate, or certify, bioethicists. Bioethics is interdisciplinary in character; its practitioners include philosophers, theologians, physicians, nurses, health policy specialists, medical humanists, and those they serve, who do not share a common background, form a guild, or follow a uniform course of study. The field embraces philosophy, law, medicine, theology, sociology, public health, literature, and the medical humanities—each with its own setting and context, whether a clinic, research institution or laboratory, government institution, or the public forum. Bioethics is beset by the reality of moral and cultural pluralism. The complex nature of the field is such that its many methods are not a surprise. What is surprising, however, is how little discussion results from this diversity.

My argument is that the multiplicity of methods reflects wider issues about the nature of ethical reflection and argument in secular, pluralistic societies. An exploration of methodology provides a further step in defining the field for itself and society. The narrow focus on method can better define the issues involved in a particular moral controversy and the focus on methodology can better define the field.

To understand bioethics and the problems of methodology, it is helpful to make a distinction between morality and ethics. Morality is concerned with practical questions, actions, and choices. It is, in the language of methodological analysis, first-order discourse. Asking what we should or can do is asking a first-order question. That is, the question stems from practical life. Ethics, on the other hand, is usually concerned with second-order discourse. That is, questions in ethical discourse are more often systematic and critical analyses of first-order practices and values. The distinction places bioethics in an unusual situation. When people think about bioethics, they are usually thinking about practical affairs (first-order questions). What should we do about physician-assisted suicide? What is our policy on abortion? Should the government control the allocation of health care resources? But the field is also involved in the second-order questions of ethics and ethical justification. It is not about "applied" ethics or about ethical theory; rather, it is about both. Bioethics lies, in a sense, between morality, as a level of first-order discourse, and ethics, a second-order discourse.

If one thinks of morality as addressing concrete questions about one's daily moral choices, bioethics has certainly sought to incorporate such choices into its field. The realm of clinical ethics or public policy decisions, for example, are concerned with moral practices and choices. At the same time, bioethics has sought to step back from particular concrete moral issues to ethics: a systematic, second-order discourse about moral choices as it does, for example, when it asks whether medical technologies can be used to enhance human life or simply to treat disabilities.

Bioethics speaks both first- and second-order discourse. The difficulty, of course, is in holding them together: the moral pluralism of the first order is often at odds with the detached reflection needed to achieve the second order, and bioethics cannot ignore either. In its consideration of assisted death or cloning, for example, bioethics needs to take specific stands on the morality of such practices. Bioethics not only addresses first-order questions; it also reflects how these claims are legitimated, which is a concern of the second order. Bioethics, as it has developed, is not simply ethics or morality. It is in both realms.

There really is no method *simpliciter* for bioethics, given this confusion of the realms of discourse and the pluralism of moral communities and moral points of view that characterizes secular societies. The moral world is complex and baroque: people articulate various facets of moral experience using a variety of concepts and moral indicators. Some describe the moral world in terms of consequences and outcomes; others, in terms of duties; still others, in terms of relationships and caring or even in terms of rights and privacy. This complexity does not per se preclude finding a method that will bring order to the anarchy. It does, however, make us unlike Descartes, who thought he could find a "pure method."

The assumption in this book is that method cannot be separated from content. The choice of method is a decision to describe and articulate moral experience in one way over another in much the same way that artists render different impressions of the same scene. This assumption about the interrelationship of method and content puts us in agreement with Wittgenstein's view that rules or methods do not exist apart from a practice or way of life. Methods, rules, and definitions are not external to a moral world view; they are its constituents.

The use of principles or cases as a method for bioethics embodies certain assumptions about the nature of the moral world. That is, a method is tied to a set of moral commitments. As the rules of a game do not exist apart from the game but are its defining element, so, too, the choice of method in bioethics becomes the rules of the game that concerns ethics and health care. If the rules change, so does the game. When the methods of bioethics shift, we may find ourselves facing very different moral issues.

One reason for the controversy in bioethical issues and concerns is that when different methods are deployed, they lead in different directions. Feminist bioethics has shown provocatively how our views of moral rationality reflect our world view. In bioethics, moral rationality is often appealed to as though it were independent of any particular stance toward reality. For example, an assumption behind most moral controversies is that moral reason operates the same for all people. However, such assumptions carry more content than is often recognized or acknowledged—and that content must come from first-order experience, which is not the same for all. Those who define moral rationality strictly in terms of consequences or outcomes will have a different view of intrinsic moral evils from those holding the view that moral evaluations are confined within certain deontological constraints. Methodological assumptions color and organize our moral experience as regularly as moral experience shapes our methodological views.

This relationship of method and substance leads to the second theme that runs throughout this book. If method is tied to assumptions about the moral world, a pluralism of methods will reflect a variety of moral views and experience. That is, the methodology of bioethics points to the reality of moral pluralism.

Many academics and intellectuals treat terms such as "moral pluralism" and "multiculturalism" as mere empty saws or buzz words; other equally serious thinkers treat these terms as holy refrains. For some, moral pluralism is not a difficult issue since they assume that a common morality can smooth over the wrinkles of plurality. These thinkers minimize differences to find similarity and common ground.

For others, moral pluralism represents a deep fragmentation and diversity that cannot be bridged. Such thinkers accentuate the fragmented character of secular bioethics. The challenge for the field as a whole is to think past an either/or position on this issue that would create a dichotomy. We need to avoid imposing a particular method on all other views, but neither can we afford to stand back in a respectful relativism refusing to make any assessments of methods or positions.

My argument is for the middle ground. The recognition of multiculturalism and moral pluralism does not, necessarily, entail moral relativism. One can hold that objective moral standards exist and still recognize that profound epistemological barriers can prevent our knowing these standards or giving expression to them in bioethics. The position espoused here is that despite its limits, secular morality can nonetheless achieve moral boundaries and avoid relativism.[4] In some ways, the radical middle ground is actually the most dangerous position; it takes seriously the need for a diversity of moral voices *and* the need to identify the common ground of moral acquaintanceship.

A fresh turn toward methodological reflection and questions can help the field move beyond these dilemmas. Indeed, methodological considerations may be the first step toward much fruitful collaboration among our various communities. We need to look for ways to acknowledge that different understandings of moral experience, different settings, and different disciplines have shaped both the development of the field and the manner in which it is currently practiced.

Again, this book is not about finding a single method for bioethics. As the book develops, readers will, I hope, discover that different methods now contribute insight, now limit our common search for consensus in this field. Bioethics goes forward in many different settings. As an enterprise, it is both individual and social; as a philosophy, it is both applied and theoretical. It occurs in research institutions, hospitals, courtrooms, and capitols—and sometimes a particular method will make stronger or weaker contributions depending on its ability to move among these various venues. My object here is to set out the conditions whereby these different methods and specialties can begin to collaborate on the moral issues in health care.

The first part of the book, "Moral Friends, Moral Strangers, and Methodology," examines the development of the field and explores the development of bioethics from a methodological point of view. Some of the most influential methods that have been used in the field are examined in these chapters. If my judgment on a particular method appears harsher than warranted, keep in mind that it is precisely the limits and possibilities in each method that will point us toward greater understanding. The clearer our grasp of a method's shortcomings, the better will be our appreciation of its positive contributions. I make the arguments from a stance that sees these different methods as complementary, and complementarity is not necessarily hierarchical and competitive. Each one captures an important part of the moral world without which our view would be fragmented and possibly incomplete.

After these methodological considerations, we will return the spotlight to the relationships of moral experience to moral commitments and methodology (part 2). The turn toward moral experience and communitarian thought (chapter 5) will challenge us to rethink our notions of community and secular society. Morality is a way of life, and moral communities are essential to its context. To share a moral culture is to share a world view in which many moral communities are at work; and this recognition constitutes the norm (or norms) for action. Such a culture makes the actions of one community intelligible to the members of another community. A community is a body of people sharing a common interest or view that is not shared by others. Communities have particular views of the good life and particular saints and heroes. Communities shape moral language so that the mean of moral terms becomes part of everyone's way of life.

This turn toward the community makes sense if one thinks of ethics as grounded in particular knowledge, as every experience must be. It also makes sense if one thinks it important to acknowledge that humankind experiences the world in a variety of ways (i.e., that the human realm is multicultural and morally pluralistic). It doesn't make sense if the emphasis on communities is interpreted relativistically— if, that is, it is used to reduce all communities to sameness. We must resist any temptation to develop *the* communitarian perspective.

At the same time, the turn toward the particular should not be used to place bioethics in a relativist position. The turn toward particular communities identifies that which can be held in common. The members of a community are, in Engelhardt's terms, "moral friends." Put another way: secular society is not merely a large community or a collection of communities that exist atomistically. Rather, secular society can be understood as many communities that have overlapping as well as different ideas, commitments, and members. Members of a secular society may be regarded in many circumstances as moral acquaintances. The category of moral acquaintances is introduced in chapter 1 but developed more fully in chapter 6. Understanding the members of secular society as moral acquaintances of one another helps explain why bioethics has had such a variety of methods and invites us to look at the moral experiences that men and women in a secular society have in common.

The procedures of bioethics, such as informed consent, prior notification, futility, or patient self-determination, are part of our common morality (see chapter 7). They are the moral practices we share in common. Bioethics must reflect systematically on the ethical assumptions that underlie and justify these common procedures. What are the moral values that underlie the procedures? What are the moral justifications for the procedures of bioethics? That reflection is its task and its natural direction for the future.

PART ONE

Moral Friends, Moral Strangers, and Methodology

1. Bioethics and Moral Acquaintances

An exploration of methodology in bioethics must confront the reality of moral pluralism. Moral pluralism is evident in the public and controversial nature of many issues in the field of bioethics. This reality has colored the field from the start. However, bioethics has had a public, controversial character not only because there is disagreement about issues but also because there is disagreement about methodology and how to analyze the issues at hand.

There are often two common responses to moral pluralism. One response is to accentuate the differences between positions and minimize any notion of common morality. A clear danger to such an approach is that it can lead into a form of relativism that it really does not matter what standards are held and the resolution of moral issues relies on the use of power. If any standard will do, if no arguments are better than others then, in the end, might will determine what is right. Another implication of this type of relativism is that there is no reason for anyone to take a position and argue for it as the best position or make that position part of one's way of life. It simply would not matter what position one took on any issue. An alternative response to moral pluralism is to minimize the differences and view them simply as "variations on a theme" as in a well-composed symphony. A problem with this approach is that it underestimates the depth of differences that exist on issues like abortion or physician-assisted suicide. This position, somewhat like the relativist one, cannot engage in argument and analysis as it does not take the differences seriously. There is a third alternative that falls somewhere in between these two. That is, take seriously the differences yet, in so doing, find points of commonality. This commonality will be fragile and tentative.

And one will have to be cautious of having points of agreement do too much work. Such agreement and commonality, I think, is best captured in the language of *moral acquaintanceship.*[1]

The category of moral acquaintanceship comes out of H. T. Engelhardt's distinction of moral friends and moral strangers. I will argue later that this conceptual scheme is incomplete. Men and women can also be understood as moral acquaintances. Moral acquaintanceship comes in a range of possible relationships. Moral acquaintances can have a "passing acquaintance" and simply understand the views of others. Or they may not only understand the views of the other but they may also share, to some degree, the moral views of the other. Some type of moral acquaintanceship is needed for a secular democracy or republic to function morally. Citizens must share a common moral view of individual liberty to morally justify the democratic or republican structures. The better we can understand moral acquaintanceship the better we can see the complementarity of methods in bioethics argued for in the introduction. The topic of moral acquaintanceship will be taken up in later chapters. In this chapter I would like to sketch out the category in relationship to bioethics, methodology, and a procedural morality.

——— THE EMERGENCE OF BIOETHICS ———

Moral questions are not something new to medicine. They did not simply emerge *de novo* in the late 1960s.[2] There is a long tradition of physician ethics that has been part of medicine's self-understanding. From the Hippocratic school in the fourth century B.C.[3] to Gregory (1724–73), physicians have sought to understand the relationship of good medical practice and morally appropriate medical practice. The work of Edmund Pellegrino and David Thomasma is a contemporary embodiment of this tradition.[4] This tradition of thought seeks to discover the moral nature of medicine in the practice of medicine. Its focus falls on the health care professional and the moral duties that are at the heart of the profession.

Physicians, however, were not the only ones concerned with the moral questions of medicine. Religious traditions have long engaged in reflection and argument on medical questions.[5] Indeed in the 1950s, 1960s, and 1970s much of the moral debate concerning issues in medicine took place within religious traditions. One thinks of the contributions of theologians such as Healy, Kelly, McCormick, Jakobovits, Fletcher, Ramsey, and others.[6] It is only in the last thirty years bioethics has emerged from physician ethics and theological ethics to become a new, unique field of inquiry.[7] In its emergence as a field of inquiry bioethics has become a field dominated by philosophers and lawyers. What caused this development?

Beginning in the 1950s, remarkable technological developments opened up whole new sets of possibilities in medicine. The era witnessed the development of medical technology such as kidney dialysis, organ transplantation, and effective and safe contraception along with safe abortions. Treatments that were once thought of as science fiction or wildly futuristic became part of the ordinary standard of care and the expectations of daily life. Such developments continue in many areas of medicine to this day. One has only to think of the advances in reproductive biology and genetic medicine to understand that the future of health care will be even more dramatic change. These are not simply medical developments. These developments affect our basic notions of family, parenting, sexuality, and society, each of which are overlaid with moral significance.

The development of medical technology—any technology—sets the stage for other moral questions, opens up possibilities, and creates choices. The creation of technology also provides a way of viewing our choices. Contemporary societies tend to view technology, for example, with a bias toward efficiency.[8] Technological developments in medicine especially prompt questions about how we should act. The developments in medicine touch the moral imagination and offer new possibilities for human life. These technologies have created opportunities at the central moments of human life—sexuality, birth, suffering, and death. These events in human life are often interpreted through moral commitments. Such events are understood within the larger framework of one's understanding of what it is to be human. Medical technology has increased our power and capacity over human life and has expanded horizons of what can be done for human beings. In turn such technological development leads to questions of what ought to be done. These questions have been central to the field of bioethics.

By itself, the development of technology in medicine, however, does not explain bioethics as it has emerged. If common assumptions about appropriate and inappropriate behavior are shared, then one could expect that moral controversies stemming from such medical technology would not have occurred. However, as the 1960s brought sweeping cultural and social change in the West they also brought different views of appropriate and inappropriate behavior. One can argue that many of these views existed before. They were not "new." However, the social changes of the 1960s gave voice to many who were once suppressed or marginalized. It was an era that acknowledged the reality of cultural and moral diversity.

One might ask why is moral pluralism such a problem for health care? It is important to recall, from the outset, that health care and medicine are not simply the applications of scientific knowledge to the human condition. Rather, health care and medicine are situated within the moral context of someone's life. Health, disease, life, death, birth, and suffering are often understood in the context of a person's moral narrative. In *Sweeney Agonistes* T. S. Eliot wrote: "birth, copulation, death. That's all the facts when you come down to brass tacks."[9] If one thinks of morality as part of one's understanding of the meaning of life, medical intervention opens new possibilities, creates options, and also raises questions. Each of these moments—birth, suffering, death, sexuality—are all open to the intervention of modern medicine. And each can be understood within a moral narrative. For example, the medical interventions that can assist infertile couples to have children can also help other, non-traditional, couples to have children. Other medical technologies have opened up the possibility of separating the unitive purpose of sexuality from the procreative. For centuries such a separation of procreation and sexual expression was beyond our control. But now medical knowledge and technology make such control and choice possible.

Moral pluralism is such a part of contemporary health care not only because medicine is moral but also because contemporary medicine requires a cooperative infrastructure in its development and delivery. Patients often find themselves being treated by health care professionals and providers who have different moral views than they do. Members of health care teams often disagree about what is appropriate or inappropriate for patients because they hold different moral views (e.g., end of life). Even in nations like the United States which have a greater free market element in health care than many other industrialized nations, there is still a significant public investment in health care. The delivery of health care is enmeshed in institutional, bureaucratic, and social structures that insure that moral pluralism cannot be ignored.

The context of contemporary medicine and health care delivery shapes the field of bioethics. The development of medical science and technology has provided patients with options and choices that have never been available. These technological developments have challenged the paternalism of a traditional physician-centered medical ethics. The public, social context of health care delivery has shifted the intellectual focus away from theological ethics to philosophical and legal ways of thinking through the issues in bioethics.

These changes, in new possibilities created by medical technology and the expressions of cultural diversity, significantly influenced the development of the field of bioethics. Social and technological developments created change in medicine and society that created a field of bioethics which moved beyond the internalism of physician ethics. At the same time, with growing cultural pluralism, Gilbert Meilaender argues that the field of bioethics focused more and more on creating and sustaining public policy. Meilaender argues that this focus on public policy has led the field to "lose its soul." That is, as the field became more and more concerned with public policy issues it became more and more captivated by the lowest possible level of discussion: proceduralism.[10]

I argue, particularly in chapter 7, that there is "more than meets the eye" in the appeals to, and the condemnations of, proceduralism in bioethics. The procedures that are adopted in a secular society, if they are justified morally, must rest on moral commitments that transcend different moral communities. The moral commitments that support procedures in bioethics are often missed by those who argue for proceduralism as well as by those who condemn it (e.g., Meilaender). Many bioethicists mourn, with Meilaender, the "reduction" of bioethics to procedural ethics or autonomy. However, the most common level of procedures and policies has more substance than is often acknowledged. There are moral assumptions that must be made in order for the procedures to be morally justified. Why have these commitments been overlooked? Why do people write of "mere procedural morality"? One can only speculate, but such an oversight may be a vestige of a Cartesian heritage in the West that assumes method to be "empty" and so it is that procedures are often viewed as "empty." So it is that secular bioethics has often searched for "more." If one assumes that methods are empty of moral content, then one may easily fail to see the moral richness within methods and common procedures.

Why start with procedural morality? One reason is that procedural morality has moral authority in the context of a secular, morally pluralistic society. A procedural morality provides a common starting point for ethical discussions. One might think of proceduralism as the first-order moral discourse of secular bioethics. Proceduralism can bridge the diversity of moral communities and procedures are some of the moral practices found in secular societies. If procedures have moral authority it is because there are common moral assumptions that underlie them. Procedures become a heuristic tool for understanding moral acquaintanceship. If procedural ethics does reflect a common moral ground, then the ground is quite fertile.

Moral common ground can be identified through procedural morality. Procedures identify the terrain of moral acquaintanceship. While procedural ethics are important for a secular society, the procedures are not as shallow as they seem. Rather, they offer a helpful way—perhaps the most helpful way—for understanding common moral ground.

—— Moral Acquaintances and Moral Pluralism ——

When one thinks of moral pluralism in bioethics one may think immediately of disagreements about specific issues that have been so much a part of the field. Abortion, physician-assisted suicide, and reproductive medicine all come to mind as points of disagreement. And rightfully so. These issues have been particularly troublesome for the field, and their problematic nature is often attributed to moral pluralism.

I think, however, that the question of moral pluralism in bioethics is understood only partially if one views it as manifesting itself in different views about specific issues. The problems of moral pluralism are manifested in methodology as well as specific topics or issues. Two people may share similar moral commitments but have very different methodological assumptions. The differences at the methodological level can lead to important disagreements in particular cases and situations or questions. For example, two people may hold that taking human life is wrong. One person holds this view under a deontological methodology while the other works from within the framework of a virtue theory. The one may view such norms as exceptionless while the other views the norms as holding "for the most part." The difference is not so much a view about a particular issue but about underlying assumptions about the moral world and what it requires. These different stances can lead to different practical conclusions about a case or about the type of public policy that should be developed. The choice of a method is not a "pure" choice that somehow takes place outside the context of one's moral commitments. Rather, method itself reflects a normative pattern about what counts and what doesn't, what is important and what is not, in moral knowing and judging. The choice of method directs, in advance, not only the activity of knowing but the content that is to be known. The two are inextricably bound together.

The introduction argued that rather than search for *the* method in bioethics or simply despair over the variety of methods, we ought to seek to understand the methods as potentially complementary. That is, insofar as methods reflect different elements of moral experience (outcomes, duties, rights, character, roles, relationships), no single method commands all the others. Rather than simplifying the moral world to one or two elements, the different methods illustrate the angles and edges of the moral world and moral experience. Choices about methodology in bioethics are further complicated by the different "settings" of the field: particular cases, institutional decisions and policy, character formation, law and public policy. There is still a further level of complication to the questions of methodology in bioethics. It is the role that the different disciplines play in the field. If the moral world of bioethics is as complex as I think it is, then we will need an approach to methodology that is sensitive to these complexities. We will need to develop a sense of the strengths and weaknesses of each method (that is the purpose of chapters 2, 3, and 4) if we are to come to a view of how they might complement one another. To further develop the complementarity of methods we will need a category such as that of moral acquaintances. This is a category I try to further develop in the later chapters in the book. For now I would simply say that the category is developed out of H. T. Engelhardt's distinction between moral friends and moral strangers.

Engelhardt has set out a distinction between moral friends and moral strangers. Those who are moral strangers share little in common save the possibility of peaceable consent. Those who are moral friends share a great deal in common. They not only share moral commitments and views but they also share common metaphysical views of the world. One might think of the abortion controversy as an example of moral friends and strangers. The clash between strangers has often come over deep metaphysical issues about the status of the fetus. Moral friends, in contrast, frequently share common views about abortion because they share deep, underlying assumptions. This does not imply that moral friends will always arrive at the same resolution. Methods in bioethics are not like formulas with variables that simply need the values plugged in. Nevertheless, moral friends share a language and a methodological framework in which they can disagree and argue.

The distinction of friends and strangers is helpful but incomplete. Part of the problem is the view of moral community that underlies the distinction (see chapter 5). While the schema of friends and strangers captures much about bioethics, much is missing. There are instances where we appear neither as friends nor strangers but as moral acquaintances. That is, there are contexts where we share some points in common and are able to articulate a common morality. It is a common morality weaker than many in the field would hope for but stronger and more robust than some think possible.

The procedure of informed consent is an example where I think moral acquaintanceship has been at work in bioethics. In the past thirty years informed consent has become an essential part of clinical practice and medical research. It is something that is supported by a wide variety of moral views. One can argue for the consent of patients from a deontological point of view, from a liberal point of view, a libertarian point of view, or a virtue or care theory. Men and women from different points of view, different traditions, deploying different methods can come to some agreement on the point of patient or research subject consent. The arguments in support of informed consent are not all the same, and they move in various directions. Nonetheless, persons holding different moral views can understand one another and share some common commitments that will support the practice of informed consent. Some may think that the category of moral acquaintance is like the language of overlapping consensus that has been the coin of the realm in many ethical circles. (Chapter 6 will examine the role of consensus in bioethics more carefully.) One advantage for the category of moral acquaintance is that it captures the fragile nature of any point of agreement in a way that the consensus language does not. The category also tries to avoid some of the problems with common morality. The appeal to common morality, so much a part of bioethics, is hampered by day-to-day moral controversies that indicate that a morality is less and less common. An appeal to common morality seems to put us in the category of moral friends. Moral acquaintanceship keeps us a bit more distant and tentative.

Informed consent also points out how limited moral acquaintanceship can be. If one thinks about the role of consent in physician-assisted suicide (PAS) one finds an example of moral acquaintanceship. Tom L. Beauchamp, for example, views the central issue of PAS as an issue of patient consent. Insofar as patients are competent, they should be able to request physician assistance in suicide. Other bioethicists hold a view that patient autonomy and consent are not sufficient to justify PAS. This view holds that there are limits to what can be requested by a patient. At this point, acquaintances become more distant. They may understand the reasons of the other but reject the reasons as sufficient.

One can see examples of moral acquaintanceship in issues surrounding removing a patient from treatment at the end of life. There can be many different reasons for why aggressive treatment is inappropriate and many different ways in which to frame these reasons. It could have to do with the wishes of the patient, or the medical indications, or views about appropriate and inappropriate use of resources. Moral acquaintances can know the other's views and positions, and agree on some points while disagreeing on others. The more controversial cases (e.g., PAS) strain the limits of agreement in moral acquaintances.

It can be argued that one might also view abortion, one of the most controversial issues in bioethics, through the category of moral acquaintanceship. There are clear and numerous issues that separate people on this issue: the moral status of the fetus, the role of the state, and the importance of a woman's choice. The differences expressed in these different views are often tied to deep metaphysical assumptions. Nonetheless, people who hold very different views can find common ground. Many who hold different foundational views can look at cases involving abortion and agree that there are moral losses in often tragic choices. Furthermore, many people, who hold very different views about the law or public policy, can agree that members of society ought to find ways to build a society that seeks to reshape the circumstances that lead to such tragic choices and provides options for women in crisis. This is the common ground of moral acquaintance-ship. The category is an attempt to develop a type of moral ecumen-ism. To be ecumenical is to look for what we hold in common.

The role of procedures can play an important key to understanding a morality of moral acquaintances. In the practice of bioethics some thinkers (e.g., Meilaender) judge appeals to procedures to be empty while others see them as the best we can hope for. I would argue a position in between. That is, the procedures have more content than many acknowledge. The procedure of informed consent, for example, embodies important assumptions about the human person, honesty, trust, and the like. It is not simply an empty ritual.

—— PROCEDURES AND MORALITY ——

Bioethical discussions often suffer from a lack of clarity around issues of objectivity, truth, pluralism, and relativism. In public policy discus-sions, people often talk as if objectivity in ethics or bioethics was analogous to objective empirical judgments about the world. At the same time, others will assume that simply because there are differ-ences we are left with relativism as the only option. Such views as-sume a modern paradigm of moral knowledge. Either moral knowl-edge is to be discovered or, if it cannot be discovered, there simply is no norm of moral knowledge.

There is an alternative reading of the situation. One can argue that objectivity in bioethics is not about correspondence or depicting things as they are. However, this position does not leave us with rela-tivism. There are standards for moral discourse that ought to be rec-ognized by anyone. How these standards are expressed (e.g., values, rules, principles, cases) will vary in different moral communities. How these commitments are articulated will depend on the views of par-ticular moral communities. Yet even if there are great differences be-tween different particular views, one can argue that there are degrees of commonality. Such common ground, in a secular society, can be found through procedures.

Different methods in bioethics have attempted to articulate the boundaries to the content of morality. Such methods begin with the universal but face the difficulty of specifying the meaning of the prin-ciple or case to a particular situation (perhaps the best known method is that of Beauchamp and Childress's principlism; see chapter 3). For example, different communities specify principles such as "justice" in different ways, or they treat beneficence and nonmaleficence as part of the same principle. One finds an analogous problem with terms like "health" and "disease." That is, while there might be a universal sense to the terms, there are certainly enormous differences in the specifications in the particular and concrete. So, given the difficulty of moving from the universal to the particular it is not clear what is gained by beginning from the universal and moving to the particular.

A universal method is not helpful for bioethics since it seems clear that ethical thought, especially bioethics, is grounded in the particular and practical (see chapters 2, 3, and 4). It seems best not to search for a universal method. Rather, it might be helpful to begin from the ground up. The turn to communities provides a way to understand the commonalities and differences between moral communities. That is, in the midst of differences between moral communities bioethics has put a greater and greater emphasis on procedures. Some claim that procedures are the best we can do for ethics in a morally pluralistic, secular society. However, it can be argued that the procedures offer more than is commonly thought. For example, procedures such as informed consent rest on common moral assumptions about honesty, truth telling, respect for persons, and trust. These are rich, complex moral notions. The procedures are not as empty as some have thought and others have mourned. Procedural morality for bioethics provides a helpful way in a secular, morally pluralistic society to identify the boundaries of morality. The procedures provide a way to identify common moral ground and the common ground provides a moral way to assess the procedures.

------ SUMMARY ------

In setting out an overview of the project this chapter has taken the view that the different methods found in secular bioethics are reflective of the moral pluralism in secular societies. The methods reflect the complexity of the moral world and the complexity of the issues in the field of bioethics. The field ought to deploy an approach to methodology that assumes that the methods can complement one another. For this to happen each method must be examined for its strengths and limitations (chapters 2, 3, and 4). Having clarity on each of the methods the field can better understand how they might work

together, no matter how fragilely. One way to understand this complementarity is to see men and women in diverse moral communities
as moral acquaintances (chapters 5, 6, and 7). A key to identifying the
common ground of moral acquaintances in bioethics is the procedural
morality that has become so central to the field. The procedures are
the moral practices of a secular society, and they embody certain moral
assumptions and commitments that provide the common ground for
a secular society.

2. Foundational Methods

One methodological approach for addressing the moral controversies in bioethics is that of foundationalism. This method actually takes many different forms in bioethics. Foundationalism includes methods as diverse as utilitarian approaches (e.g., Peter Singer) and virtue theory (e.g., Pellegrino and Thomasma). In the literature some authors refer to foundational methods as "theoretical." However, in discussing this method it is important to be clear about what is meant by "theory" as theory can describe other approaches in bioethics.

Stanley Clarke and Evan Simpson cite two particularly prominent understandings of "theory" in moral philosophy. One is a rationalistic formulation in which theory "requires a set of normative principles governing all rational beings and providing a dependable procedure for reaching definitive moral judgments and decisions."[1] In this view, principles provide the foundation for judgment and serve to justify judgments. Foundational methods are usually this type of theory. The other approach is to construe ethical theory in terms of a process of reflective equilibrium that tests ethical beliefs against other moral and nonmoral beliefs with the aim of developing a set of beliefs that fit together. Moral judgments are made in a way that renders them coherent with the larger set of one's moral beliefs. Methods such as casuistry or principlism are this second type of theory. The process of reflectively balancing considered judgments is usually set in a social context as part of the development of a general social contract. These two very different approaches to moral theory illustrate the deep ambiguities that exist in defining the notion of "moral theory." Furthermore, these problems with "theory" are not unique to bioethics. They are often found in all fields of the arts and sciences.

So, how does one understand a foundationalist method? Foundational methods argue for a fundamental starting point that centers and directs the method. A foundational starting point in bioethics sets out a comprehensive picture of the moral world of medicine and health care. A foundationalist method claims that the starting point, whatever it may be, is the *correct* starting point for ethical deliberations. From this starting point foundational methods develop an integrated body of principles or rules.

What is it that makes the foundational method appealing? There are, perhaps, many reasons for its attractiveness, but one can argue that a primary attraction for the foundational method is the hope of developing the most definitive solutions possible for practical ethical issues in bioethics or other areas of ethics. In making practical choices in particular circumstances, we look for ways to justify the choices that have been made. When there are disagreements about moral issues we try to find a basis for agreement or for justifying the positions that we hold. One of the best ways to resolve dilemmas is to obtain agreement at the most basic, foundational level of assumptions. If there is a basic agreement at the foundational level there will be some control of the discussion about particular moral questions. The appeal of the foundational method lies in securing the most basic, general agreement.

In its basic appeal, however, can be found the seeds of important difficulties for the foundational method. The very strength of the method, its general and universal basis, becomes its weakness. The problem is that moral questions, particularly those of bioethics, exist in the concrete and the particular not in the universal and the abstract. Foundational theories provide a difficult starting point for bioethics insofar as the field needs specified content to address the particular issues of the field. In the realm of secular discussions, the foundational method will be limited to the extent that there is moral pluralism. That is, because of their generalities, foundational ethical theories often have gaps between the theory and the issues of practical ethics (bioethics). To bridge these gaps the theory must be specified in some way. Often this specification happens through the generation of principles or rules or virtues. However, in specifying the theory the power of generalized agreement is jeopardized as the theory becomes more particularized. For example, a theory might begin with a foundational principle that to act ethically one must "do good and avoid evil," or a theory of justice might begin by requiring that being just requires rendering to "each his due." While such principles may generate agreement they are not very helpful as they do not tell us what good should be done, what evil avoided, or what is due. Foundations need content if they are to address practical moral questions. However, if one assumes that moral pluralism and cultural diversity are important and not to be overlooked, content will be open to variation. There will be different particular views that shape the content of the theory. So every foundational theory needs content, but with content often comes the loss of general agreement.

Content, however, is not something that is added to a foundation. The pluralism and diversity of moral experience raise another important question for any foundationalist method. Each ethical theory not only assumes content but also assumes a form (e.g., consequentialist, rule, or duty based) to shape and express the content of the moral vision. As stated previously, methods reflect a particular moral view. Methods do not stand outside a particular moral view but are one expression of a view. Each theory articulates a description of the moral world. We know, however, that there are many different descriptions that are possible as is evidenced by the different foundational accounts. Each theory in bioethics must make particular background assumptions about moral values, moral theory, and moral reason in order to identify and resolve moral controversies. When such commitments are shared, moral controversies can be identified and addressed. When such commitments are not shared, it becomes more difficult to deploy the method of moral theory.[2] The dilemma posed by moral pluralism is not merely that there are too many foundational theories. If that were the case, one could still hope to discover some way to choose among them. But as one compares the different theories one comes to recognize that they represent not just different moral points of view but they also conceptualize fundamentally different understandings of "reason" and "theory" as well as the compass of "moral."

One will find in the field of bioethics a recurrent acknowledgment of the differences at the metaethical level. There are differences about metaphysical questions and the grounding of ethical claims. There are also differences about epistemological questions as to how the content of the moral world is known (e.g., moral sense, intuition). Some have thought that the differences at this level are not important for the resolution of practical bioethical questions. Robert Veatch argues, for example, that there is a convergence of the different metaethical positions.[3] Others, like Beauchamp and Childress, hold for a view of convergence in common morality. I will argue, in the end, for a view that takes the differences seriously but looks for complementarity where possible. The best starting point for complementarity is procedural ethics.

This chapter will examine some of the important foundational methods that have been used in bioethics. It is not intended to be simply a destructive chapter that says "all theories are bad." Rather, theories help to articulate the deepest assumptions of a moral world view and can help clarify points of agreement and disagreement that will be important later in the book. This chapter will articulate some of the problems with each method and with the foundationalist methodologies as a whole. But the focus on the limits is *not* intended to imply that the foundational methods are useless nor is it meant to imply that the best view will have to be problem free. Rather, the limits are highlighted so that we might discover common ground from which to build. Understanding the limits helps to understand the strengths and positive contribution of each method.

Foundational methods often begin with some element of moral choice. For example, Singer begins with attention to consequences. Alan Donagan turns to duties and obligations. Pellegrino and Thomasma look to character. Each starting point founds the theory that is developed. Implicit in the choice of each starting point are assumptions about the nature of moral reason.[4] But part of the postmodern dilemma, made clear by much of the literature of feminism and multiculturalism, is that reason is not a univocal concept. It is a mutable concept. The choice of a founding starting point is complicated by the complex nature of moral choice. A phenomenology of moral choice reveals that there are many dimensions to moral choice: consequences, obligations, promises, relationships, and character are but a few.

Theories are purchased at a price of accepting particular metaphysical and epistemological commitments as well as moral assumptions.[5] Pellegrino and Thomasma think there is a *nature* to medicine that can be known and serve to direct moral deliberations. Singer's choice of consequences gives a greater latitude for moral choices than a foundational method that begins with deontological commitments. Such choices have very *practical* consequences. For example, if one takes consequences to be central to moral justification, this choice of a starting point, in itself, reflects a particular moral understanding that in turn structures the interpretation and resolution of particular cases. If someone began reflection on a moral controversy with another moral appeal as primary, this would engage a different set of moral values and commitments. For example, a person who understands certain acts to be intrinsically evil (e.g., the act of suicide) will describe and judge cases differently from a consequentialist. A person's moral values and beliefs influence what will and will not count as "rational" or "irrational."

Even if agreement existed on which foundational appeal should be deployed, that agreement would not, in itself, be sufficient to resolve a moral controversy in the midst of moral diversity. Each theory needs a content and needs a method by which it can be applied to specific moral cases and dilemmas. For example, if we agree that consequences should direct our moral choices, some standard is then required to evaluate the consequences. What standard should be used? Should we look for the greatest good for all or should we look to increase the greatest good of the average? For example, in reforming the health care system of a nation should bioethicists argue to improve the outcomes for all members of society, or the poorest members, or simply the average of all? Each assumption will yield very different schemes for reform of the system. Or, if an appeal is made to a hypothetical choice theory such as some of the hypothetical contractor accounts, each account will require a particular moral sense, or set of values, to direct the choices of the contractors.

This is the case even with theories that build in mechanisms for ranking values. Deontological theories, for example, often establish "absolute obligations" which can never be violated. Some contract theorists, such as Robert Veatch, provide a ranking of contracts and principles. Other types of foundational theory, such as natural law accounts, have "basic goods" which cannot be acted against and which therefore require rules that govern choices (e.g., the principle of double effect).

Each choice of a foundational model and moral content particularizes a moral theory. The price of such specification is the loss of the theory's generality. This creates something of a "Catch-22" situation insofar as the foundational method, which seeks universality, requires particularity. Put another way, the value commitments that foundational methods must make in order to resolve moral controversies in bioethics engender the moral controversies and contested points which they were to resolve.

The particular choices and controversies in bioethics can be a heuristic device for moral philosophy. Because bioethics addresses concrete cases and issues it must address moral questions in a way that general philosophical ethics can avoid. Bioethics makes it clear that if one is to address concrete moral questions, such as whether or not postmenopausal women ought to be free to bear children, then one must have a particular ranking of moral values. Bioethics is forced to deal with the complexities of morality and moralities in a way that general ethics can escape.

Bioethics makes it evident that there is an abundance of answers to the foundational questions. There is no way to choose among them without begging the foundational moral questions at issue. This claim is not simply empirical or sociological. It is an epistemological claim that, in a secular context[6] open to moral pluralism, there is no way to know the foundational criteria by which to resolve content-full moral controversies. In scientific research we are able to judge the success of a paradigm or theory in light of its ability to account for particular data. In ethics it is often the "data" of morality that are in dispute.

However, the claim should not be made stronger than it is. It is not a metaphysical claim. It is not a statement about the existence of moral truth. It is, rather, simply a claim that we do not know, in general secular terms, how to articulate rationally the means of discovering the moral truth. In later chapters I will argue that what we know, in the moral realm, is shaped by what we believe about the world.[7]

In what follows I will show the foundational diversity of some of the major theories of contemporary bioethics: the utilitarianism of Singer; the deontology of Donagan; the contractarianism of Veatch and Norman Daniels; the natural law theory of John Finnis, J. Boyle, and Germaine Grisez; and the virtue theory of Pellegrino and Thomasma.[8]

───── A SURVEY OF FOUNDATIONAL THEORIES IN BIOETHICS ─────

The foundational methods in bioethics cover several different approaches to the field; they are foundational because of their universal and grounded characteristics and attendant difficulties. This section is intended to survey the different foundational methods, present what they do, and understand the limits of their applicability to bioethics. It is worth repeating that pointing out the limitations is not meant to be a dismissive argument. Every method has limits. However, understanding the limits enables us to better understand the contributions of each method and how a method might be used.

SINGER'S UTILITARIAN PROJECT

An Overview

Utilitarianism has played an influential role in shaping modern Anglo-American moral philosophy[9] and bioethics. A good example of its influence is found in the work of Peter Singer.[10] Singer begins by outlining the parameters of what ethics is not. It is not a set of prohibitions about sex, nor is it an ideal system about "goodness" unrelated to the difficulties of human life, nor a set of religious claims, nor, finally, is it relative and subjective.[11] Singer then attempts to set out the minimal structure of a moral theory that transcends particularities. He argues that every variety of moral theory makes some appeal to a *universal* element.[12] By "universal" Singer means that an ethical principle cannot be justified in terms of any particular or sectional group. Ethics takes a universal point of view.[13] Singer holds that the universal element of ethical thought leads one to take a utilitarian position.

However, one must do more than simply describe ethics as "universal" in order to secure a workable account of ethics or bioethics, for this description by itself is formal and empty. Singer goes on to argue that any moral theory must appeal to the *interests* of those involved. In arguing that the ethical point of view must be universal, Singer argues that it is a universal balancing of the interests of agents. The choice of an action, and its justification, must include an account of all interests and the effect that choices will have on those interests.[14] A crucial part of Singer's project is to determine what counts as an "interest." He argues that to have an interest one must have a capacity for suffering and enjoying.[15] From a moral point of view one's concern for one's own interests must be extended to include the interests of others. Singer thinks this universal calculation of interests is the basic starting point for moral theory, and this is why he holds a utilitarian view of bioethics. He argues that utilitarianism captures this basic structure of ethical theory and to go beyond this stance and accept nonutilitarian moral rules or ideals should happen only if there are reasons for such additional commitments.[16]

Singer argues that a central element built into the ethical enterprise is a notion of "equality."[17] He understands equality to be a basic ethical principle and not an assertion of fact. This claim, he acknowledges, is not based on any actual equality that living beings share.[18] According to Singer the basic principle of equality is "the principle of equal consideration of interest."[19] That is, we give equal consideration in our moral deliberations to the like interests of all affected by our actions and decisions. Singer sees this as a minimal principle of equality in that it does not dictate equal treatment for all, but merely equal consideration of everyone's interests. Health care needs, for example, ought to receive equal consideration. Such consideration does not mean, necessarily, that all will receive the same treatment as there may be other factors such as age or other morbidities that must be factored into the judgment. Nonetheless, the interests of all should receive equal consideration.

But the simplicity of Singer's methodology subsumes complex questions that need to be resolved for the theory to work. For it to work agreement must be reached as to whom or what counts as an interest bearer and on what counts as an interest. The theory must be able to resolve the difficulty of calculating how competing interests between agents should be calculated.

Pluralism of Foundations and Moral Values

Singer's project is open to two criticisms when faced with moral pluralism: the pluralism of models of reason and the pluralism of content. One can argue that Singer's project too narrowly construes our moral "experience." Singer argues that any method of ethics must be, in some sense, a *universal* appeal.[20] This appeal to universality is a reflection of the aspirations of modernity. Other moral appeals, such as those to universal laws, ideal observers, categorical imperatives, or human rights attempt to capture a universal dimension in moral choice. These are but a few ways the universal dimension can be captured in different methodologies. Singer posits the most basic universal element is "interests" and he argues that the most basic element of the moral world is a universal respect for interests. He argues that even appeals to "rights" and "duties" begin in a "universal" appeal. That is, such appeals are understood to apply to all moral agents. In Singer's view, however, moral appeals to "rights" and "duties" go beyond the minimal foundation of "interests." Those who would develop a theory of rights and duties are obliged, in his view, to offer an argument justifying why we should go beyond the minimum theoretical structure.

A difficulty here is that the "universal" component of moral theory does not always provide the common denominator for which Singer argues. Singer's use of the "universal" in terms of interests and a deontologist's use of the concept in terms of moral maxims represent two very different views of the moral world. For example, it may be in the universal interest of a terminal patient to end his life while his physician (a deontologist or virtue-oriented individual) may have a universal interest not to kill. In linking the concept of universality to the concept of interests Singer has given the concept of universal a particular meaning.

The different possible interpretations of "universal" illustrate a difficulty for moral language in that the same word can be used with very different moral force. This problem with language and meaning is a difficulty that occurs again and again in bioethics. The ambiguous use of language is one reason why I use the term of moral acquaintanceship. In many cases, when the same moral term is used, but with different nuances and interpretations, we see that people are not moral strangers for they share something in common. Nor are they moral friends, as they have important differences that separate them. Bioethics is a collaborative enterprise in which there are conflicts of interests between individuals, institutions, and often within individuals or institutions.

The different meanings to "universality" and "interest" reveal that this utilitarian method will need a particular content. Singer, like any utilitarian or consequentialist thinker, must build a particular moral sense into his theory. Without some ranking of interests, a theory cannot deliver common resolutions to particular moral controversies. This is especially the case in health care which is often a cooperative venture involving people with competing sets of interests who become involved in moral controversies. For example, if one considers making decisions about the level and type of treatment to be given a persistently vegetative state patient, how does one weigh equally the interests of the patient, the family, other patients, those affected by the consumption of resources, and the insurers who pay for the use of resources without some ranking of values? There is no objective point of comparison from which to evaluate the added time "alive" for the patient, which may be valued by some members of the family, against the use of resources which may be of importance to providers, other members of the family, or other patients. There is no yardstick or metric by which to evaluate the different interests and decide which is of greater value, unless the theory devises some way to rank them. Unless the theory gives some ranking of values, we are asked to compare interests without a point of comparison. Another dimension of the calculation problem is that it may undercut "personal" moral commitments by assuming a certain disinterested stance toward one's interests. Singer would regard this a virtue. Still, it reflects a particular moral commitment which may not be shared by others. How is one to determine, when one is the only person affected, which action to take? In some way the sense of "interest" will need to be more clearly defined and specified.

If Singer's theory is to direct common moral choices then certain assumptions must be made. The first assumption is that moral reason is about consequences and outcomes. Second, there must be some common assumption about whom or what is an interest bearer. Third, there must be some kind of assumption about the nature of interests. Fourth, the theory must determine how competing interests between interest bearers are to be calculated. Each of these assumptions, however, further particularizes the theory. When moral agents do not share the same assumptions about moral reason or the ranking of moral values the theory will be useless in resolving moral controversies in morally pluralistic societies.

DONAGAN'S *THEORY OF MORALITY*

An Overview

Alan Donagan's foundational account of morality stands in sharp contrast with Singer's utilitarian approach. Donagan sets out a philosophical theory based on the common morality of the Judeo-Christian tradition.[21] In this account the nature of common morality is deontological. There are a number of perfect duties that structure moral deliberation. The structure of his account is monistic and deductive. He begins with one basic principle from which he deduces secondary principles. This is not surprising since he argues that a *theory* of morality is "a theory of a system of laws or precepts, binding upon rational creatures as such, the content of which is ascertainable by human reason."[22]

Donagan contends that morality is founded in the nature of reason and that the moral requirements of reason are best captured in the Judeo-Christian tradition. He hopes that the method he develops will not depend on any theistic beliefs. Donagan argues that the common morality of Judaism and Christianity is a theory of a system of laws and precepts binding upon rational creatures *as such*.[23] His accent here upon a discernable rational character of this morality is likely an inadequate account of Orthodox Jewish understandings of morality, is surely a false presentation of Orthodox Catholic moralities, and fails to appreciate the radical differences amongst the moralities of the various Christian religions. The theory he develops will be a deductive system based on precepts contained in a first principle which can be known by reason. Such a system is binding on all rational agents and it would be irrational not to follow it. However, to establish his theory, Donagan makes crucial presuppositions about nature, the human person, and human action.

Donagan enumerates the crucial presuppositions of the tradition as he understands them. First, the human being, considered as a moral agent, is a rational animal. Second, the moral world human beings inhabit is a system of nature in which events occur according to morally neutral laws. Donagan then argues that traditional common morality holds that the relations of rational creatures with themselves and one another are derivable from a first principle. While it has had different formulations, Donagan takes the first principle of common morality to be the Golden Rule: Do not do to your fellows what you hate to have done to you.[24] Donagan cites the appearance of the principle in other, non-Western cultures (e.g., Confucius) as evidence that the rule is rational and not culturally based. He understands common morality to be a system of laws and precepts about human actions when actions are considered *objectively*.[25] The most familiar form of these laws and precepts are prohibitory in form. He defines "precepts" as any universal proposition of common morality.[26] These precepts of common morality, he thinks, can be deduced from a basic principle.

Donagan acknowledges that there are other formulations of the Golden Rule. He examines the formulation of the Golden Rule in the Thomistic school. In Germaine Grisez's formulation, following Aquinas, one should "Act so that the fundamental human goods, whether in your own person or in that of another, are promoted as may be possible, and under no circumstances violated."[27] However, Donagan holds that a Kantian formulation is simpler and more inclusive than the Thomistic formulation. One should "act so that you treat humanity, whether in your own person or in that of another, always as an end and never as a means only."[28]

From this first principle Donagan moves to develop specificatory premises which enable him to deduce the precepts of his system. He needs, as any monistic system does, some way to specify the relationship of the fundamental principle to particular actions. That is, he needs these premises to identify species of action as falling, or not falling, under the fundamental principle.[29] The specificatory premises play a crucial role in moving from the general principle to particular judgments. According to Donagan the movement from the general principle to particular judgments is as follows:

1. An action is wrong if it treats human beings as a means only and fails to respect as an end.
2. Action X is treating a human being as a means.
3. X is wrong.

The second premise is the specificatory one in that it delineates a species of action that treats human beings as a means only. From the specificatory premises Donagan deduces first-order precepts which further specify duties and ways of acting toward oneself, others, and society.

One might think of an action such as stealing, murder, or lying as actions which treat human beings as means. It is this starting point that supports the development of precepts against stealing or murder. The crucial difficulty is to identify the source of these premises and how we know them to be true. Here one might note the fundamentally different approaches taken to the problem of truth telling in Orthodox Catholicism verses Roman Catholic Catholicism. Latin Catholicism, under the influence of St. Augustine, has held an absolute prohibition against lying. One of the foundational influences on Augustine is his identification of truth with God. The Orthodox churches, however, have allowed the possibility for "therapeutic deception." [30] The development of the specificatory premises is crucial if the method is to address particular moral dilemmas. However, the development of such premises, as illustrated by the rules about lying, can move the same method in very different directions.

Donagan compares the derivation of the specificatory premises to the process of legal reasoning which works from precedent cases to explain new cases. It appears, however, that the specificatory premises are more conclusionary than operative; that is, they reflect the conclusions of common morality we want to support rather than function as part of a process for reaching those conclusions. Donagan appears, in the end, to be more of an intuitionist than a deductivist.

Pluralism of Foundations and Values

A fundamental problem for Donagan's method is that he relies on "common morality" as the normative substance of his theory. In so doing Donagan builds into his theory a particular moral world view with a whole set of claims and values. The difficulties of an appeal to "common morality" are at least twofold. First there is the question of whether a common morality exists.[31] Many have argued against it.[32] What remains are bits and shards of a once powerful and commanding moral vision. There are family resemblances among the utterances of various moral languages but no coherent, single language. Indeed the challenge of postmodernity, illustrated in bioethics, is the challenge of moral pluralism. Many of the ongoing debates about the practices of abortion or physician-assisted suicide are painful reminders of the absence of a common morality. There is indeed the question whether Christians share a common morality.[33] Donagan assumes a robust content-full view of common morality. On this account many of the debates in bioethics raise serious questions about the assumption of a common morality. In appeals to common morality, one finds some of the difficulties facing secular bioethics. If the common morality can be articulated in such different fashions, it is hard to find what is common. Beauchamp and Childress (see chapter 3) assume a much more general sense of common morality. This weaker sense allows them to explain the controversies in bioethics as debates over the specification of a moral general sense of common morality.

A second problem raised by claims of a common morality is the question of why such morality should have normative force. One might argue against Donagan and others that the same common morality to which he appeals has, in the past, sanctioned practices that people today often find morally repugnant (e.g., hereditary slavery, segregation) and prohibited practices many now accept (e.g., abortion). The questions about common morality are a recurrent problem for the different methodologies in bioethics (see, for example, chapter 3 on principlism).

Beyond the questions about the content and role of a common morality one might raise a question about the view of moral rationality found in either of these approaches. Certainly Singer, as well as others like Albert Jonsen and Stephen Toulmin, would dispute Donagan's idea of the "rational." Singer, in contrast with Donagan, views the rational as an evaluation of consequences. Indeed Jonsen and Toulmin argue that the model of deductivist reason used by Donagan is a misconception of moral rationality. They argue that moral reason must be understood as *practical* rather than theoretical.

A further question for Donagan is that the different systems make crucial assumptions about the relationships of duties, principles, consequences, and human action, which are controversial. Like other deontological theories, they reflect an ordered moral world with a clear delineation of responsibility. While this approach is agent centered, it nonetheless misses dimensions of agency such as the virtues or intentions. Also, unlike much common moral practice, there seems to be no significant attention to the consequences of moral actions. These appeals to reason, however, presuppose a certain foundational, universalist conception of "reason" which then justifies the process of reasoning which leads to the content. Indeed, Singer would object to limiting morality to "rational beings." Just as Singer's choice of "interest" gives content to his theory, so Donagan's choice of "rational" gives content to his method. Again, the choice of a moral "appeal" contains within it the framework for the choice of a particular content.

The development of the premises seems to incorporate more of the conclusions Donagan wishes to achieve than he would like to admit. For example, he states that the development of the specificatory premises is analogous to legal casuistry. This means, however, that one will need to assume a great deal more about moral cases and the content of the theory than the deductive process would lead one to believe. Legal casuistry works within a defined process, tradition, and content.[34] Legal casuistry is practiced within the boundaries of legislative codes, acknowledged precedents, as well as publicly acknowledged judicial offices that have the authority to settle disputes. When applying this analogy of legal reasoning to moral reasoning, one finds crucial elements lacking in the moral realm. It is not clear what the proper content of the premises should be, or who is definitively to interpret the content of the premises. It also is not clear whether the content actually comes from the deductive process or whether the theory merely summarizes the norms which are already in place. For example, in resolving some of the questions of reproductive medicine one has to make assumptions about the moral status of the fertilized eggs. One does not deduce such specific premises from the general ones but simply makes ad hoc assumptions in particular cases.

The specificatory premises reflect a ranking of moral commitments without which the theory would be unable to guide moral choices and judgment: a significant conceptual gap exists between the general principle on the one hand and particular problems and controversies on the other. Like other moral theories that are monistic and deductive in structure, the first principle has a general character.[35] Left alone, such first principles are not content-less, but they are so general that they are unable to guide particular judgments since they can be interpreted in so many ways. Without some type of specification, one will never be able to frame a range from which to choose. Donagan needs something like the specificatory premises to enable the theory to address particular moral problems. With such premises the theoretical stance might effectively limit the range of choices in particular problems. However, without specification, the theory remains at such a general level that it is powerless to address moral questions so as to direct particular conclusions. The conceptual difficulty is, however, to present an argument as to why we should accept one set of specifications over another. The theory becomes particularized by appeal to the specificatory premises and the content of these premises is justified by an appeal to common morality.

NATURAL LAW

An Overview

There are many interpretations of the natural law tradition in ethics, political philosophy, and bioethics.[36] The natural law tradition has a long, influential history in Western moral thought going back to the Stoics[37] and continuing to the present. It has had a profound influence in Western culture less for its content and more in terms of its presuppositions about reason and morality. Because of its assumptions about the ability of reason to know moral truth, natural law theory has inspired many of the hopes of modern moral philosophy. The tradition embodies the belief that human reason can discover the moral law embedded in human nature which transcends time and culture. The common assumption of natural law theories is that moral duties can be known by reflection on nature. However, natural law traditions have differed significantly in their views about which interpretation of nature should be normative and, as a result, they differ about the content of the moral theory derived from reflection on human nature. They also differ on the normative level and function of the theory.

Natural law approaches have at least two elements which make them attractive. First they assume that moral knowledge (principles) can be apprehended by reason. If the project succeeds there would be a way by which moral agents, from different cultures, would be bound together. Second, they take seriously the pluralism of moral reason by trying to include a variety of arguments (deontological, teleological). Again, if successful a natural law theory would circumvent some of the foundational issues about the nature of moral reason.

There are two significant conceptual issues, however, that confront any theory of natural law. Such a theory must address the difficulty of arguing that any appeal to "nature" should take priority over other moral appeals. When nature is understood as the product of chance, mutation, genetic drift, and evolution, it is not clear why it should be considered morally normative. Second, any account will have difficulties in terms of content in specifying what "nature" is. Singer, for example, would be critical of a natural law theory's appeal to *human* nature as the basis of morality. Even those who appeal to the natural law differ in what is regarded as morally normative. Some appeal to biological structures while others appeal to rational structures. The fact that there are numerous theories of natural law, some at great variance with one another, indicates that moral truth may not be as easily apprehended as proponents have asserted. Like other deductive moral theories, natural law theory needs content. The first principle of the natural law must be specified if it is to guide moral subsequent judgments. Thomas Aquinas, for example, argued that the first principle ("Do good and avoid evil") was specified in secondary principles.[38]

A contemporary example of natural law reasoning in bioethics is found in the work of Finnis, Boyle, and Grisez.[39] Finnis, Boyle, and

Grisez's discussion of the natural law is set out in the context of their examination of the justification for nuclear deterrence. Like Donagan, they accept the norms of common morality as the data to be explained by moral theory. The theory they propose differs from consequentialism and other teleological, goal-directed theories, as well as from Kantian, deontological, duty-oriented theories which seek to ground moral norms in the *nature* of morality.[40] They hold that their theory possesses the strength of a teleological account through a theory of human flourishing, while avoiding the weakness of teleological accounts which cannot support any deontological constraints which can protect individuals. Natural law has the strength of deontology in that moral judgments are grounded in the rational nature of the moral subject. Yet it avoids deontology's overemphasis on universalizability which Finnis, Boyle, and Grisez take to be but one aspect of common morality. They understand the natural law theory as combining the strengths of both and avoiding the weakness of both.

Finnis, Boyle, and Grisez hold that morality can be derived from the human good, that is, "the goods of real people living in the world of experience."[41] A "good" is any object of interest while a basic human good is something essential or necessary for full-being as a human. Out of these reflections they formulate the first principle of the natural law system: "One ought to choose and otherwise will those and only those possibilities whose willing is compatible with integral human fulfillment."[42] In speaking of human fulfillment they go to great lengths to point out that they are speaking of the good of all persons and communities. They are not deploying a contemporary view of "self fulfillment" as the satisfaction of desires. This latter view of self-fulfillment is characteristic of MacIntyre's "cosmopolitan"—the rootless modern who has no commitments to a parochial community and its vision.[43] While the cosmopolitan carries no commitment to others, the Finnis, Boyle, and Grisez notion of human fulfillment is tied to the fulfillment and good of others. Finnis, Boyle, and Grisez hold that there are five basic human goods: life, knowledge, aesthetic experience, excellence in work and play, and relations. These are different aspects of full human flourishing and each constitutes a principle of practical reasoning.

Like Donagan, the natural law authors have to make their general principle less abstract in order to address day-to-day moral issues. They propose three intermediate principles to help shape the interpretation and implementation of the first principle. The three are:

(1) the Golden Rule, (2) a principle which excludes hostile feelings, and (3) the principle that one should not do evil that good may come of it.[44] These intermediate principles shape the rational prescription of the first principle into definite responsibilities.[45]

From these intermediate principles, which outline prohibitions of actions and the way actions may be done, specific norms can and are deduced.[46] In this tradition there is a strong emphasis on both actions which are prohibited (*mala in se*) and the intentions and decisions of the agent. The agent's relationship to action is a central part of their theory. Like Donagan, Finnis, Boyle, and Grisez develop a theory of action. Unlike Donagan, however, the role of "intention" plays a crucial role in how they propose we evaluate actions.

Pluralism of Foundations and Values

Again, as with the other theories, we encounter the two crucial conceptual difficulties confronting any moral theory. Their interpretation of the natural law is centered on an understanding of basic human goods. The theory does try to offset some of the foundational problems by incorporating different moral appeals (e.g., ends, duties, intentions). However, it does rely, most heavily, on a deontological structure; that is, one can never act directly against the basic human goods. In this account the structure of the natural law is founded on the basic human goods of nature. These goods are to be protected and promoted.

There is a crucial foundational question of why someone should take nature to be normative. There are those who would argue that nature is not the uniform, law-governed reality the natural law tradition presupposes. Rather, nature may be seen as the outcome of impersonal forces and chance. However, even if nature is the law-governed moral reality portrayed by the theory, one may still ask why nature should be morally normative.[47] Simply because nature *is* a certain way does not mean that we *ought* to act in certain ways.

Even if one accepts the normativity of nature, the natural law theorists still face the second conceptual problem of determining which description of nature should be held as normative. "Nature" is subject to many different descriptions. The multiplicity of descriptions is problematic for a natural law theory since the description is taken to be morally normative. In reflecting on human nature, for example, Thomas Hobbes believed human beings to be motivated by their desire for pleasure and aversion to pain. David Hume held that there was a common human nature centered on moral sentiments.

Kant, in contrast, held the view that human beings were moral agents in their rationality. The force of morality is not at all connected with human desires. The contrast of these philosophers illustrates how difficult it is to determine what constitutes human nature.

Even within Roman Catholicism's culture of natural law there are at least two different descriptions of nature. On the one hand "nature" is often "read" from the physical structures of the world. That is, the moral norms are inferred from claims about physical or biological characteristics of human beings. This physicalist tradition has framed many of the moral rules about sexual conduct (e.g., teachings on artificial contraception, in vitro fertilization, and masturbation). This reading of the natural law is perhaps best exemplified by Pius XI's teaching on procreation in *Casti conubii* in which he says that "any use of matrimony which deliberately frustrates the natural power of matrimony to generate life is an offense against the law of God and of nature."[48] The moral norms of sexual intercourse in marriage are determined by the biological functions of the act. There is, however, a "person"-centered natural law tradition which focuses on the rational nature of the human person as the moral norm of nature. It takes into account the psychological aspects of human behavior.[49] Finnis, Boyle, and Grisez offer a third alternative for understanding human nature within this tradition. They deny that their system of basic goods is inferred from empirical or metaphysical claims about the biological or psychological composition of human nature. Rather, these basic goods are "self-evident" though not innate. Still, others have criticized Finnis, Boyle, and Grisez for offering a "Kantian" account of the natural law and failing to situate the duties of the moral life within a proper teleological understanding.

The contrast of these different accounts illustrates the difficulty of defining "nature" in a common, content-full way. These are deep conceptual problems since the description of nature that is used is, of course, crucial to the judgments that are made. The description provides the ranking of values which enables the theory to resolve moral controversies. For even if one could establish the foundation of the theoretical appeal to be morally normative, the first principle of the theory will be empty without a description of nature which articulated a set of values like the basic human goods.

Natural law theory then faces at least two basic questions. First, while attempting to incorporate a number of moral appeals it is not clear why the appeal to nature should be adopted. Second, there is the problem of moral content; that is, one needs a moral point of view in order to interpret nature. Any account of "nature" is a particular account which builds in certain views of moral reason and certain rankings of moral values. There is also a third issue that confronts this foundational tradition. It is the question of the *function* of the natural law. Some natural law theorists understand it to function on a meta-theoretical level or to be supracritical of moral norms,[50] while others understand the natural law to have a much more practical, normative function. The issue for bioethics, when a natural law methodology is used, is where the foundation for the method is built.

CONTRACTARIANISM

An Overview

Contractarianism in bioethics is an adaptation of the tradition of social contract theory of Western political philosophy. This tradition in itself is quite diverse[51] and that diversity is reflected in the application of contract theory to bioethics. One approach to social contract theory has been to justify the contract by appealing to hypothetical contractors and the principles to which they would agree. This is the method of justification deployed by Thomas Hobbes, John Rawls, and Norman Daniels in bioethics. Another contractual approach to justification has been an appeal to tacit consent. Perhaps the most outstanding example of a tacit consent theorist is John Locke. Locke argues that by submitting to some structure, moral agents tacitly give their consent to it. In bioethics the work of Robert Veatch, which appeals to foundational contracts, is a model of tacit consent. In other ways, however, Veatch uses an implicit appeal to hypothetical contractors.

Veatch sets out his case by first arguing against the tradition of physician paternalism in medical ethics.[52] Veatch regards the Hippocratic tradition as one that attempts to benefit the patient according to the physician's ability and judgment. Veatch argues that this approach is unacceptable since it is grounded only in the agreement of a professional group and does not include the patient. According to Veatch there is no *reason* for anyone outside the group to conform to physicians' judgments about what benefits patients.

He attacks the traditional Hippocratic ethic in two ways. First, Veatch argues that to the extent that, in the Hippocratic tradition, the professions invent the norms, a fundamental mistake is made. It assumes that a given profession has the inside track on knowing what the moral norms are. There is no reason to make this assumption. Indeed, the profession may even be biased. The mistake is to assume that the hypothetical professional consensus could be substituted for the hypothetical consensus of society. Veatch also attacks the content of the Hippocratic tradition. He argues that the norm for physician conduct (to benefit the patient) is inadequate as a moral justification in contemporary medical practice. The norm is inadequate as it is constructed only from the physician's perspective and it looks only to the benefits and harms as assessed by physicians and sidesteps any deontological constraints such as truth telling.

To address questions of medical ethics, Veatch argues that one must understand the physician–patient relationship within a wider social context. One can conceive of this relationship through a model constituted by three contracts. The model of triple contracts can be utilized by those who understand morality as objective and absolute or those who conceive morality as subjective.[53] According to Veatch the "objectivist" attempts to determine the ethical principles by use of reason or discover them by appeal to moral sentiments. The "subjectivist" sees people coming together for the practical purpose of harmonious survival. Veatch thinks that regardless of the perspective one takes there will be no difference in the principles that result. He believes that either project, done from *the moral point of view,* will lead to principles to which both objectivist and subjectivist can agree.[54]

The first contract is a basic contract which specifies the content of the ethical system for all society. The basic contract is summed up in a number of principles: beneficence, autonomy, honesty, avoiding killing, justice, and contract keeping.[55] This is the set of principles that hypothetically ideal contractors would favor. The key commitment is to the method of the basic contract. The second contract exists between society and the different professional groups. It captures the relationship of the profession to society as a whole; that is, there are role-specific duties which arise not from within the profession but from social expectations of the profession. Veatch argues that this second contract is symbolized in the process of licensure.[56]

The first two contracts fix the broad boundaries of the patient-physician relationship, which is the third contract. In this contract, questions such as the extent of care are negotiated.[57] Veatch does not appeal explicitly to hypothetical contractors to justify the basic social contract or the contract between physicians and society. The justification seems to take the form of some type of historical consent. Since there is no process of real consent, Veatch must rely on a form of tacit consent. By submitting to the structures, people give their consent. However, implicitly Veatch has a hypothetical methodology in mind. On his account a proper account of the basic norms and principles would be articulated by ideal contractors. In the real world we come as close as possible by asking real observers of the moral order to reach an agreement on norms. Diversity, of course, within the group, is essential.

A crucial conceptual issue, inherent in the basic contract, is the relationship of the principles to one another. Veatch acknowledges that there may be apparent conflicts between the principles. The problem of conflicts between principles has been well known in bioethics and in the method of principlism (see chapter 3). However, Veatch holds that a balance can be struck. He rejects the possibility that there can be an overarching principle which governs all.

Veatch sees beneficence as the best candidate, but he rejects this candidate since production of the most good may be unjust.[58] Since he thinks the search for a single principle is likely to prove fruitless Veatch raises the possibility of giving some lexical order to the principles. He argues that if the nonconsequentialist principles are to have any power they must be given a priority over the principle of beneficence.[59] When there is a conflict between any of the nonconsequentialist principles one should seek that course of action which produces the lesser violation of the principles on balance.[60]

Another version of contractarianism is found in the work of Norman Daniels.[61] Daniels deploys the method of hypothetical contractors to address questions concerning the allocation of health care resources. In elaborating his position Daniels relies heavily on the work of John Rawls.[62] Daniels includes health care institutions among the basic institutions of society. The concern of the hypothetical contractors is the structure of social institutions. He argues that health is a special good and that health care institutions should be governed by the principle of fair and equal opportunity. He argues that prudent contractors will limit access to health care in light of other goods and the needs of other institutions. Of course, the successful use of this account will depend on shared views of what it is to be a "prudent" contractor. Access will be restricted according to what the contractors think people should rationally want at a particular age or situation.

Issues of Foundations and the Ranking of Values

As with the other foundational methods the choice of a starting point and the ranking of values are at issue for the contractarian approach. Veatch's contractual theory, for example, must develop a set of moral principles that gives content to the method. In this way his theory appears to include a variety of moral appeals; that is, there are appeals to both deontological and consequentialist elements of morality. Unlike the other theories, the contract method can incorporate a number of elements of the moral action (e.g., consequences, first principles, basic goods, or intentions) since it appeals to a *method* rather than a set of values or principles. It is a method that appeals to the way in which the moral discussion is constituted.

The contractarian approach seems to incorporate a number of different understandings under the rubric of "agreement." However, the difficulty with the foundation becomes evident as soon as one turns to the content (the ranking of principles) within the contracts. For example, Veatch thinks he can undercut the theoretical differences of the "objectivists" and "subjectivists." Success, of course, would depend on the capacity of the model to set out content acceptable to both. An objectivist may reject a ranking of principles as wrong when the contractors have not included certain moral features. Objectivists, like Finnis, Boyle, and Grisez offer a very different set of principles which, they hold, cannot be ranked.

Even if one accepts Veatch's model of contracts, it seems clear that, through his treatment of beneficence, he has built the structure with a deontological commitment. He restricts beneficence because he is afraid it may be used to wrong someone even if it maximizes the good. Within bioethics one also finds appeals to other sets of principles such as the four principles of Beauchamp and Childress.[63] One could easily imagine rational people wanting to include principles such as the "sanctity of life" or a principled commitment to equality of treatment. The rankings of principles and values are, indeed, the crucial problem in a morally pluralistic society for there is no shared framework which would allow men and women to choose one set or ranking over another and come out with universal agreement.

Daniels's model faces similar difficulties with his appeal to "the rational." Utilitarians, like Singer, would object to such a move as "speciest" and limits the moral realm to the rational and human. Perhaps the second conceptual issue, that of a particular ranking of moral principles, is the most difficult criticism for Daniels to meet in that, *ab initio,* he must presuppose a particular view of "rationality."

The commitment to particular moral values is hidden in his discussion of "health." Health is a concept that is not exhausted by an empirical reality. It is also a concept with personal and moral significance.[64] What constitutes health for one person may be different from how it is constituted by another. For example, for one person correcting a deviated septum may be a matter of health while for another simply a matter of cosmetic surgery, or two patients with the same diagnosis and of similar age may make very different choices about treatment. Health is a good that is understood within one's view of the good life and the good death. The allocation of health care resources will be meaningless unless one assumes a standard of health and its relationship to other goods. Daniels's contractors, then, will need some ranking of values or principles if they are to make the kinds of choices Daniels hopes they will make about health care.

In the use of the contract method one finds yet again the difficulty of determining how we can possibly justify particular moral judgments. Such judgments are crucial for bioethics as a field. Veatch's theory relies on both a particular set of principles and on a way of balancing between them. But it is far from clear how the balancing is to take place. At the very least, there is no metric by which to judge or balance in practical choices. Indeed in describing the development of a set of principles and their balancing, Veatch speaks of these acts as taking place in "the moral point of view."[65]

Contractarian theories trade on the foundational ambiguities of form and content. They base their content on some particular understanding of the rational which articulates a moral point of view. Yet bioethics is like a marketplace filled with different vendors selling their "moral point of view." The epistemological difficulty, however, is our inability to discover which moral point of view is the correct one. Without this, it is unclear why Veatch's, Daniels's, or anyone else's list of principles should be accepted.

The contractarian approach can accommodate the pluralism of moral appeals by incorporating a variety of principles within a contract. But it is not clear why one should accept any one particular set of principles or their ordering. So the approach is not successful in meeting this pluralism of appeals. Also, this approach must rely on a moral sense or ranking of values to develop content for the contract. Different contractors might well develop very different types of contracts for bioethics.

An Overview

Another method for resolving moral conflicts in health care has been an attempt to deploy virtue theory as a method for bioethics. Virtue theory has an ancient basis in Western culture in the writings of Plato, Aristotle, and the Stoics. Aristotle saw the virtues set within the context of a teleological view of the world in which the virtues, both moral and intellectual, were habits of the intellect and will that enabled men and women to fulfill proper ends.

Similarly Edmund Pellegrino and David Thomasma[66] offer an approach to bioethics which begins by describing the practice of medicine as a virtue. This is a method that they have developed extensively over the years. Given the different types of moral theory—foundational and nonfoundational—one might think of virtue theory in the nonfoundational category. Indeed, it may well be. One could assume a set of virtues required for certain social practices (e.g., medicine) and develop them. However, the Pellegrino and Thomasma virtue method, when understood in the context of their work in the philosophy of medicine, is a foundationalist methodology. In their work on the philosophy of medicine they argue that there is a "nature" to medicine that makes it what it is; medicine is not only a social practice constructed by a society (that can be constructed in a number of ways). Their account of virtues relies on these realist assumptions about the nature of medicine.

To summarize the conceptual framework of the method, it is best to return to their earliest philosophical work. They begin by defining "[M]edicine as a disciplined body of knowledge is a science respecting the perfection of lived bodies concertized by skill in experiencing and effecting connections between corporeal symptoms and remedies."[67] The practice of medicine is an art, a science, and a virtue (of practical reasons) which aims at achieving the end of a right and good healing action for the patient.[68] Medicine involves right reasons about things to be done.[69] Pellegrino and Thomasma understand medicine as a practical science defined by a proper end (health) that guides the physician to act in certain ways on behalf of the patient.[70] The focus of medical ethics, understood through the practice of medicine, is the professional code of the physician with a focus on the proper healing, and good of, the patient.[71] In their later work they clarify the relationship of the beneficence of the physician to the autonomy of the patient.[72] In a more recent development of their thought they continue to speak of virtue as a character trait which breeds a disposition to live in accord with the moral law. These virtuous traits are defined by the ends and purposes of human life.[73]

The method of virtue ethics relies on views of what people should be like and the types of practices that ought to be supported. The practices reinforce the virtues and the virtues reinforce the practices. To know the virtues one has to assume that there are certain practices that are normative. The virtues are defined in relationship to those practices. Thus, implicit in the appeal to virtues is a view of the types of practices a society wants to encourage. For example, a society which wants to maintain a traditional Western family structure might promote the virtue of fidelity and punish infidelity. It is in this view of promoting certain "practices" that Pellegrino and Thomasma argue that medicine is a practical discipline.[74] They try to develop a set of virtues required for the successful practice of the discipline of medicine. This concept of building the virtues around the notion of a practice ties the appeal to a notion of human flourishing. They argue that to flourish as human beings certain practices are necessary and those practices require certain virtues. As such the method of virtue theory is a very different approach than other methods in bioethics. It is less focused on the boundaries of behavior and it is more about what health care professionals (and patients) should be like. Rather than look at the limits of which actions are permissible and which are not, the method seeks to articulate a vision of whom the health care professional is called to be.

Foundations and Ranking Moral Commitments

Like the other theories examined in this chapter, many would argue that a method of virtues fails to sufficiently capture the whole spectrum of people's bioethical concerns. For example, some would argue that an appeal to virtue theory does not adequately address concerns about consequences or appeals to autonomy. While the appeal may address the virtuous practitioner, it cannot adequately address the autonomy of the patient. One can well imagine a moral controversy created between a virtuous person whose pursuit of virtue intruded on the autonomy of another. For example, a physician might pursue the treatment of a patient even when the cost of the treatment may deny medical care to others. Or one could easily imagine that a deontologist may see the moral universe in terms of actions that ought never be done (*mala in se*) no matter the virtue being pursued. Such a view will not primarily be concerned with the virtues of an agent, or a profession, or the virtues of certain acts. Another problem is that it is not always clear what the foundation of the appeal actually is. Is the appeal to virtuous acts or virtuous persons? The method makes a crucial assumption that there is an *essential moral nature to medicine* which, once known, is the end and guide for professional behavior.

Even if we could agree that one set of virtues ought to be morally normative, it is not clear that the virtues alone would help make moral choices. A given situation might call upon an agent to do two things each of which can be described as virtuous. For example, one might understand a situation to call for the virtue of "honesty" while also calling for "temperance" and "compassion." The virtues would seem to require the assistance of other appeals (e.g., rules, role models) to guide their application. There also needs to be some way to rank and balance the virtues in relationship to one another.

A second conceptual difficulty is brought out by the fact that controversies in bioethics frequently involve many actors such as patients, physicians, families, and institutions. The virtue theory of Pellegrino and Thomasma is too narrowly rooted in a physician's point of view and does not adequately account for the virtues and vices of others. This method of the virtues only addresses the virtues of other health care professions or the role of patients when subordinated to physicians. In turn one would need some tradition which addressed how the different sets of virtues interacted. Medicine is a social and cooperative venture, and it is not clear why the physician's interpretation of the patient's good should be the practice which dominates the field. The more one understands "health" as a moral good the more one must understand that any understanding of health must be rooted within the particular value structure of the patient. One critical difficulty for the virtue theorist is to show why one view of the patient's benefit (the physician's) trumps others.

A virtue method also assumes that one can generate a coherent and accepted professional ethic which can serve as the cornerstone of the theory. Alasdair MacIntyre has argued that virtues are defined by the

practices they support. The conceptual problem for an appeal to virtues in bioethics is that there are differing views about *the practice of
medicine.* This is made clear in the controversy over physician–assisted
suicide. Each understanding, each "profession," will have its own
creed of the profession and list of professional virtues. Perhaps the
most difficult problem for a virtue methodology is whether or not
there is one set of virtues for physicians. The contemporary discussion
of the physician's role in assisted suicide, for example, reflects some
widely different understandings of the medical profession, the patient's good, and the virtues of the physician. Some might see assisted
suicide as serving the good of the patient in that it alleviates suffering
while others may see the preservation of life as a virtuous act of courage. This is certainly the position taken by physicians like Timothy
Quill.[75] Diverse views of the virtues reflect the difficulty of defining
"practices" in a secular, pluralistic culture.[76] It is not clear that all
would, or should, share the same conception of medicine and health
so as to allow the development of a virtue ethic. Finally, even if one
could articulate a virtue theory for the profession, it is not clear why
it should be *morally* normative in the profession or outside of it. Even
if we focus on physician–oriented virtues, closure eludes us.

Virtue theorists, such as Pellegrino, Thomasma, and MacIntyre, do
highlight important themes for understanding the moral life and
moral community. This is a theme taken up in the second part of this
book. Appeals to moral virtues present an alternative to the normal
ways in which methodology is discussed in the circle of professional
ethicists insofar as it develops a set of reflections on the good life and
how the good life should be lived. It is, as Aristotle says, what the good
man does. On the one hand a virtue theory is limited as a method in
bioethics because of its particularity. Virtues are specific to a community. On the other hand a virtue theory is successful because it directs
our reflections toward understanding the moral life within a community. Virtue methodology directs out attention to the fact that medical
choices are made within the moral narratives of individuals. Precisely
because virtue theory is tied to communities and practices, I will return to virtue theory in chapter 5. Virtue theory plays a dual role.
Within a community the virtues help members to understand who
they are to become. Virtues are central to the life of particular communities. Nevertheless, we can discover some common ground within
secular societies. The common *practices,* such as free and informed con-

sent, reflect a commitment to certain virtues such as honesty. Virtue theorists remind us of the importance of practices as sources for reflection in bioethics. They remind us that practices rely on underlying moral commitments. And so it is that practices can help establish where there is common ground in secular, morally pluralistic societies.

THE LIMITS OF FOUNDATIONS: FORM AND MATTER

There are at least two conceptual issues that confront every attempt to use a foundational method to resolve moral controversies in the secular, morally pluralistic context of bioethics. These problems can be labeled as issues of form and matter. Each foundational method must make some assumptions about the structure (form) of moral reason. Each foundational theory gives some order to the different possible elements in moral choice and justification (e.g., consequences, obligations, duties, ends, agreements, virtues). The conceptual dilemma is that there is no view from nowhere. Each choice about form reflects a particular view of the moral world. This chapter is testimony to the many different assumptions and choices built into the foundational models. To argue that one form of reason surpasses another presupposes a logically prior understanding of moral reason. But this begs the question.[77] Methodological assumptions (e.g., the nature of moral reason) are expressions of commitments and views of the moral world. Methodology does not stand apart from content.

Even if one could resolve the problem of form one would still be left with the conceptual issues surrounding the content (matter) of a foundational method. Each method requires some set of moral values, and some ranking of values, if the method is to resolve moral controversies. Without a content the theory remains empty and impotent.[78] This problem is particularly troublesome for a foundational method in bioethics insofar as the field defines itself by resolving moral controversies in health care. To achieve such resolutions a method will need a particular content.

Furthermore in a morally pluralistic society, with a diversity of views of human life and moral "goodness," there is no reason to think that the content of a theory will, or should, constitute the set of moral commitments for all agents.[79] Unless a set of moral commitments is shared, however, neither the structure of the theory nor the solutions it develops will provide rationally convincing resolutions of moral controversies for those who do not share a particular theoretical view. In this we have a positive direction for understanding bioethics.

Foundational methods in bioethics, unlike theories in moral philosophy generally, have a stake in developing particular, content-full solutions to moral controversies. Reaching such solutions touches the very heart of the work of bioethics.[80] The field has been marketed as providing content-full resolutions to moral controversies in a secular world. However, to address particular controversies one has to supplement the general, foundational level with specifications or stipulations that work as bridges between the general and the particular. In order to develop such "bridges," one needs to build into a moral theory a "thick" set of descriptions and stipulations. Such moves, however, clearly particularize a theory, which seems at odds with the philosophical hopes of secular bioethics. One strength of the foundational method—its universality—is also its weakness. The universal needs content to address bioethical questions. But with the content comes particularization.

Historically one finds a good example of such a thick theory in the practice of Roman Catholic moral theology in the High Middle Ages and Renaissance. This, of course, is but one example and not the exemplar. Within the general practice of moral theology is a combination of elements such as rules, principles, paradigmatic cases, and theory. Many of the paradigmatic cases, moral principles, rules, and virtues were established in the Christian culture long before the full-scale development of the natural law in moral theology.[81] These different elements all worked together to guide moral choices and evaluate moral justifications. The Christian culture also provided structural elements (e.g., the sacrament of penance) to guide people through the maze of moral values, insights, and practical problems. This is not to say that this culture was monolithic. Indeed, there was a pluralism within it. However, cultural and political power were used to maintain a particular moral culture. It developed an important role for authority structures in defining the moral life and, in this way, held together the different interpretations of Christianity.

In its historical evolution the distinctions between these different elements became blurred so that in the twentieth century one finds a moral appeal based on a faith in reason with little mention of religious faith.[82] The highly rationalist model of the natural law is as convincing as any other rational theory. Separated from the context of cases, rules, saints, and virtues, the theory is open to wide interpretation and unable to address particular moral issues. In the literature of bioethics one finds an interesting phenomenon going on as well. One might call it expansionism in methods. If one looks at some of the different methods (e.g., Gert-Culver-Clouser or Beauchamp and Childress) one sees over time an "expansion" of the method to include other appeals and methods as well. Such developments reflect the complex and diverse ways through which the moral world can be understood and expressed.

The history of moral thought in the West reveals a profound yet subtle shift from the hegemony of one moral vision to a pluralism of moral visions. It is this shift toward moral pluralism that has set the stage for the postmodern dilemma in ethics, a dilemma made evident in bioethics. This shift raises important questions about the moral authority of the state and society. In the High Middle Ages one finds a dominant moral culture represented in the Christian church and enforced by the state. However, with the secularization of the West, following the Reformation, moral thought has gradually become "multicultural" and pluralistic. The power of the state to enforce a moral culture has gradually been restricted. This pluralism often provokes a tension in secular societies.

The differences among the foundational methods leads us to search anew for how common morality can be understood. What is the common morality and how might it be best described? I think that a model of moral ecumenism and acquaintanceship are most helpful. I want to use the language of acquaintanceship rather than the language of consensus as appeals to consensus can be misleading as they seem to promise more than can be delivered (see chapter 5). The common ground of moral acquaintances can be identified through the proceduralism that is so much a part of bioethics.

One can think of the example of procedures for gaining consent from patients or research subjects as a lens through which to see moral acquaintanceship in bioethics. Procedures like informed consent and refusal or prior notification have become central to the practice of medicine and bioethics. The procedures illustrate a point of acquaintanceship among the different foundational methods. While they work from different perspectives each of the foundational methods support such procedures. Nevertheless, there are important differences that emerge. For example, Singer's appeals to interests would support

expanding the practice of consent to areas like physician-assisted suicide while the Pellegrino and Thomasma view of the nature of medicine and virtues would not support such a development. Understanding the different foundational methods allows us to understand
not only what the differences are but why there are differences.

Some have sought to resolve moral dilemmas by sidestepping these
metaethical problems confronting foundational methods by raising
the important question of how much agreement does a method need
for bioethics. One way is to drop the use of foundational theory and
guide the resolution of moral controversies by appeal to middle-level
principles.[83] Such appeals build in a shared content and attempt to
avoid any particular structural appeal. Another way to approach the
problems presented in this diptych is to address the problem of the
specification of principles to particular moral controversies by working with particular moral cases.[84] Recent literature in bioethics has
sought to recast moral philosophy in terms of the practice of casuistry as a way to avoid the problems of theoretical models and respond
to the dilemmas of applied ethics. The challenge for these alternatives is whether they can resolve moral controversies without attending to the foundational question about the constitution of the moral
world and the specification and ranking of values. Both the appeals to
middle-level principles and to casuistry, however, assume some form
of common morality. Perhaps the more modest claim would be that
they hold that men and women in secular health care share enough in
common in their moral values, language, and reason to resolve moral
controversies. These alternatives reflect the effort to develop a method
in bioethics that will resolve moral controversies with resolutions that
will bind men and women generally.

3. Ecumenism in Bioethics:
The Appeal to Middle-Level Principles

The field of bioethics has been defined, in many respects, by the work of Tom L. Beauchamp and James Childress. In their book, *The Principles of Biomedical Ethics,* they argue that a method using four middle-level principles can resolve moral controversies in bioethics.[1] One can argue that Beauchamp and Childress were among the first to explore the terrain of moral acquaintanceship in that they understood both the foundational differences in bioethics and that there are overlapping points of view. Since the first edition of *Principles* Beauchamp and Childress have developed considerably the principlism method. This chapter will highlight some of these developments but it will examine, primarily, the fourth edition of *Principles.*

The method, simply stated, holds that there are four principles that articulate the necessary conditions of common morality for health care and bioethics. These four principles provide the common ground for biomedical ethics, and they can be brought to bear on moral choices through a process of *balancing, weighing,* and *specification.* This process relies on a *criterion of coherence* to guide the balancing, weighing, and specification.

Beauchamp and Childress hold that by focusing on the balancing and weighing of the principles, the dilemmas of the foundational methodologies outlined in chapter 2 can be avoided.[2] The principles are like ecumenical statements that find common ground in areas where only differences have previously been seen. In this way the principles potentially resolve two problems. First, the principles allow bioethics to sidestep the difficulties that confront the foundational methodologies by articulating a common morality for bioethics. Second, the principles enable bioethics to address, and resolve, particular moral issues in the field.

As with many ecumenists' statements, Beauchamp and Childress seek to downplay their own foundational differences and highlight what they describe as "the morality we hold in common." Later, I will argue for an ecumenism that begins with differences and searches for common ground. In the end, I think the latter approach to ecumenism will lead to a more realistic understanding of agreement. In the fourth edition of their book Beauchamp and Childress explicitly discuss their assumption of a "common morality." This is an assumption that was clear, but not explicit, in earlier editions of their work. Their principlism method relies on an assumption of common morality.

To the extent that the principlism method relies on an articulation of what men and women hold in common, the principles are like ecumenical statements. But ecumenical statements must always be read with caution. They must be read for what they say and for what they leave unsaid. On the one hand, the authors of such statements can hold very similar views and so the differences they bridge may be very shallow and the statements in fact actually achieve very little. On the other hand, the differences can be quite profound and yet the desire for commonality can be so strong that the bridge may be quite weak. Once again the statements achieve very little. Beauchamp and Childress seem to make the point that they are describing and, as necessary, reconstructing a common worldview.[3] However, when their accounts are used by others, who have less in common, the principles can become a *source* of problems rather than a solution to them. A crucial question in assessing the principlism method in bioethics, as well as for the methodology of casuistry discussed in the next chapter, is how to gauge the strength of common morality and what role, if any, it should play in bioethics. While Beauchamp and Childress assume the existence of a common morality, others, such as H. T. Engelhardt,[4] argue that there is only a very weak sense of common morality and that the primary source of morality is that of particular communities. In a sense I am arguing for a middle ground: there is a common morality that is less robust than many assume but more vibrant than Engelhardt concludes, and procedures are a fruitful way to identify the common ground.

One can ask as well if the principles are the best expression of common morality. In many ways the principles face problems similar to the foundational methods. The principles need to be related, in some way, to the particular dilemmas of bioethics. Beauchamp and Childress, in the development of their method, have relied more and more on the processes of specification, weighing, balancing, and coherence to relate the principles to particular bioethical cases. However, these processes carry moral commitments within them. They are not applied, like a recipe, in the same way by everyone. In a morally pluralistic society, these processes are likely to introduce as much disagreement as agreement.

The first section of this chapter will give an overview of the most recent version of the principles offered by Beauchamp and Childress. In recent years there have been numerous criticisms of this project and so the second section will summarize some of those criticisms. But the "standard" criticisms outlined in the second section have missed the mark. The criticisms of Bernard Gert, Danner Clouser, and Ronald Greene, for example, rely on a foundationalist methodology. It will be argued that an understanding of communitarian thought gives a more comprehensive understanding of both the contributions and limitations of the principlism method.

The third section of the chapter argues that the fragmentation and pluralism of moral and cultural views can make it difficult to recognize even what a moral principle is. The issues of "form and matter" from chapter 2 are revisited in this chapter. "Principles" are understood as part of a reasoning process. To deploy a method of principlism is to make certain assumptions about moral rationality and justification. Such are controversial assumptions. Intuitionists, casuists, and feminists, for example, have all challenged the "principlism" model of moral justification.

——— THE METHOD OF MIDDLE-LEVEL PRINCIPLES ———

In the fourth edition of their book Beauchamp and Childress explicitly develop their account of moral justification. They understand morality to be "social conventions about right and wrong human conduct that are so widely shared that they form a stable (although usually incomplete) communal consensus."[5] They clearly assume that there is a common morality and that it ought to be the starting point for moral reflection.

Beauchamp and Childress argue that in justifying a moral choice one attempts to show sufficient reason for one's claim.[6] To justify a particular judgment one engages in a process of reason-giving which may lead to the level of moral theory. Justification for moral judgments can take the following pattern:

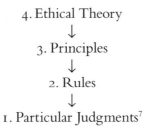

4. Ethical Theory
↓
3. Principles
↓
2. Rules
↓
1. Particular Judgments[7]

They go on to discuss three different models for justification that can be deployed. The first model focuses on moral theory and justification. It is a model of deduction that they call "the covering precept model."[8] This is a deductivist method that holds that moral judgments are deduced from a pre-existing theoretical structure of normative precepts. Moral judgments are justified in this model if and only if they can be inferred from the theory. They argue that there is a weakness of this model in that it fails to capture the richness and detail of moral experience. Further, this method opens one to an infinite regress of justification. Finally, perhaps most important for understanding their project, Beauchamp and Childress state that this model assumes that there is one correct theory. This assumption leads to the foundational controversies discussed in the second chapter. They argue that such an assumption does not need to be made in order to address moral dilemmas in bioethics. They argue that "little is lost in practical moral decision making by dispensing with general moral theories."[9]

The second model of justification is that of inductivism or "the individual-case model."[10] Induction assumes that a society's "moral views are not justified by an ahistorical examination of the logic of moral discourse or by some theory of rationality, but rather by an embedded moral tradition and set of procedures that permit new developments."[11] The inductivists hold that we learn by experience about how we should handle more problems. For the inductivist moral rules and guidelines are developed in light of particulars of cases. Beauchamp and Childress are sympathetic to this model of moral reasoning but they think it is an inadequate understanding of moral justification since there are many occasions when different judgments are reached in similar cases. They argue that "[i]nductivist theory, then, needs to be buttressed by an account of the proper role of rules and principles in adjudicating disputes" in particular moral judgment.[12] They do not think that a method of justification will eliminate ambiguity, pluralism, or differences at the level of specification of particular concrete choices.

Beauchamp and Childress adopt a third model of justification: *coherentism*.[13] This method is associated with John Rawls's use of reflective equilibrium in *A Theory of Justice*. This use of coherentism, along with Henry Richardson's process of specification, is an important development in the principlism method that is not found in earlier editions. They understand coherentism as moving in *both* top-down and bottom-up directions. The goal of the method is to bring together moral commitments, intuitions, rules, and principles into a coherent relationship. Justification for any particular judgment depends on the mutual support of all aspects. Beauchamp and Childress argue, however, that coherence itself is not sufficient to account for the moral life. The Pirates' Creed of Ethics, while coherent, is "morally unsatisfactory" for Beauchamp and Childress.[14] Ethics must start with considered judgments, drawn from moral experience. Justification must begin with a set of moral commitments that are accepted without recourse to other judgments. Such a set of commitments forms a sufficient foundation for bioethics. Indeed, it is as much of a foundation as can be hoped for. The process of justification must remain faithful to the principles and concepts that provide the starting point for the account.[15] A central assumption of the coherence method is that as we achieve greater and greater coherence the best explanation is that the beliefs are justified and should be accepted. The goal of moral thinking and justification is to start with the most carefully developed set of considered judgments and to continually increase coherence.

In light of this coherence method Beauchamp and Childress then turn to the four middle-level principles. They argue that "four clusters of moral principles" are central to biomedical ethics and the practice of health care.[16] These principles are understood as general guides "that leave considerable room for judgment in specific cases and provide substantive guidance for the development of more detailed rules and principles."[17]

What is the advantage for using the principlist approach over the foundational accounts discussed in chapter 2? Beauchamp and Childress argue that one can have theory without foundations. They write that: "Our presentation of the principles—together with arguments to show the coherence of these principles with other aspects of the moral life, such as the moral emotions, virtues, and rights—*constitutes* the theory in the present volume. This web of norms and arguments *is* the theory. There is no single unifying principle or concept, no description of the highest good, and the like."[18]

The method of the middle-level principles is advanced as an attempt to develop a normative ethics for biomedical problems in a pluralistic context which cuts through the foundationalist difficulties. The principles, in this method, are understood to provide action guides that can be shared across theoretical perspectives.[19] However, Beauchamp and Childress also understand that they are seeking normative principles in a pluralistic context. The appeal to middle-level principles is an attempt to address the problems of applied ethics in contexts that often have many different senses of what is morally appropriate.

It is important to note that in the most recent edition of their work Beauchamp and Childress also give a high profile to the role of "common morality" in moral justification. Indeed they understand common morality as the source of the principles.[20] This is a crucial evolution in their project as it seeks to offset the problems raised by moral pluralism. Any attempt to understand and evaluate the principlism method must also involve some assessment of "common morality." Is there one? Are there variations? How do these variations affect the way principles are understood and used? Should we assume that common morality *ought* to be normative?

If the appeal to middle-level principles is to provide a *modus vivendi* for the justification of moral judgments in a pluralistic context then the extent to which it succeeds will depend on some shared view of the interpretation and application of the principles. In controversies in which people may stand within the same moral framework and share the same models of moral reasons, the principles may be of great help. But the Beauchamp and Childress method is presented to address controversies in the situations where the background assumptions are not *necessarily* shared. The principlism methodology assumes that common morality forms the framework for the principles. Differences in judgment emerge as people specify and balance the principles. Others in bioethics, most notably H. T. Engelhardt, Jr., argue that the differences emerge because we lack a common moral framework.

Beauchamp and Childress characterize the principles as *prima facie* binding; that is, all else being equal the principles bind. However, they are not absolutely binding. Each principle has weight (due consideration) which should be given to it but does not have priority over any other principle. This view of principlism has led to one of the most repeated criticisms of this method. Some have argued that the method fails to articulate how, lacking a prioritization of the principles, we are to determine which principles are to be followed under what circumstances. Beauchamp and Childress, however, do spell out four requirements for the infringement of prima facie binding principles. First, one must consider whether or not the moral objective has a chance of success. Second, the infringement on a principle must be the last resort. Third, the decision must represent the least possible infringement. Finally, we must seek to minimize the untoward effects of such infringements.[21] At the same time the method seeks to bring together the principles through a process of specification, weighing, and balancing.

The task of professional ethics is to address particular practical problems. There are at least three crucial assumptions that must be made for a middle-level principle approach to succeed in applied ethics. The first assumption is that the middle-level principles can generate action guides that can be specified to, and fit, the context of concrete situations. The second assumption is that there is a common morality that ought to be normative. The third assumption is that middle-level principles can be developed that cut across boundaries set by moral theories and ways of living. Thus, we are able to come to general agreement. An appeal to these principles should be sufficient to offer a justification for moral judgments that would be satisfactory to others regardless of which theoretical perspective one holds. The goal of a principle-driven account of moral judgment is to make normative moral judgments while avoiding the quagmire of developing a comprehensive, foundational moral theory. The method hopes to avoid a relativistic account of moral judgment as well as the difficulties of adjudicating the claims of different moral theories that have dominated discussions of normative ethics in the modern age.

BEAUCHAMP AND CHILDRESS

In developing their four middle-level principles of justice, autonomy, beneficence, and nonmaleficence Beauchamp and Childress seek to establish a basic method in which to talk about ethics and bioethics. They understand the use of the middle-level principles as a way to offset the pluralism of moral theories. They write: ". . . we stand to learn from all these theories. Where one theory is weak in accounting for some part of the moral life, another is often strong."[22] The development of the appeal to middle-level principles as a method in bioethics allows us to take advantage of the insights of different theories without coming to theoretical agreement or the development of a metatheory that ties all the other normative theories together.

Beauchamp and Childress also try to navigate between understandings of principles as either 'absolute' or as 'situational' and 'summary'. They see rules and principles as prima facie "binding but not absolutely binding."[23] As such, they argue for their principles as prima facie binding. The heart of the method is the use of the specifying and balancing of the principles to achieve and maintain coherence. They stress the importance of specification and coherence in the fourth edition of their work. They note that abstract principles have "insufficient content to resolve" the moral dilemmas that arise.[24] The content is developed in the process of specification.

The first principle, respect for autonomy, involves a number of possible ways to be specified.[25] The principle is stated as broadly as possible. The principle of respect for autonomy is the "personal rule of the self that is free from both controlling interferences by others and from personal limitations that prevent meaningful choice, such as inadequate understanding."[26] Unlike other philosophical presentations of autonomy, the principle set forth by Beauchamp and Childress does not focus on the autonomous 'person' but rather on the autonomous *action*. In this they seek to avoid intractable discussions of defining 'person' or ideals of autonomous choosers.[27] They turn to 'respect for autonomy' and 'the exercise of autonomy' "which refer to normal choosers and their choices rather than ideal choosers."[28] In delineating how normal choosers act autonomously they set out three criteria. Autonomous actions are those undertaken intentionally, with understanding, and without controlling influences that determine the action.[29] One can argue that this method may not avoid the conceptual difficulties involved in defining 'person' as the concept of 'action', which implicitly involves many of the controversial elements of 'person'.

Beauchamp and Childress's second principle is that of nonmaleficence. Unlike other principlists they distinguish nonmaleficence from beneficence. They argue that the two need to be separated in order to capture distinctions of ordinary moral discourse.[30] If one tries to fit them under one principle, Beauchamp and Childress hold, one will be forced to distinguish them in some way or another.[31]

The principle of nonmaleficence is: "[o]ne ought not to inflict evil or harm."[32] Crucial to applying the principle is knowing what counts as a 'harm'. They define harms to be " . . . thwarting, defeating or setting back of the interests of one party by the invasive actions of another party." In defining nonmaleficence they argue that one must distinguish harms that are wrong and unjustified from those that are not.[33]

The principle of nonmaleficence supports, like the other principles in turn, several moral rules such as: "Don't kill"; "Don't cause pain"; and "Don't disable"—all of which are prima facie binding (that is, if one undertakes an action that causes such harm one must offer a moral justification for the action). Beauchamp and Childress use this principle to outline what they see as 'due care' in medicine.[34] Under this rubric the authors address other controversial topics of biomedical ethics such as the obligations to treat and the principle of double effect, which they interpret as a rule of nonmaleficence. As the principle of autonomy protected the competent person, the principle of nonmaleficence protects the interests of both competent and incompetent patients.

The concept of beneficence includes "acts of mercy, kindness, and charity."[35] The authors understand beneficent action to include all forms of action intended to benefit other persons. The principle refers to the moral obligation to help others further their important and legitimate interests.[36] Already under the one principle of beneficence we have two possible interpretations. One requires the provision of positive benefits and the other requires the balancing of benefits and harms (utility).[37] There is also the second 'essential addition' to the principle of positive beneficence: The limits of the principle are hard to determine since it is not clear whether it enjoins us to promote what is good as well as prevent what is bad.[38] The danger is that the principle might be interpreted as a morass of obligations that exceeds definable limits.[39]

The authors consider an example of someone who sees a drowning man while walking past a lake. While the passerby is not required, if there is undue risk, to save the drowning person, the passerby is required to call a lifeguard or the police. To establish limits to the obligation of beneficence Beauchamp and Childress set out five criteria:

1. Y is at risk of significant loss or danger.
2. X's action is needed to prevent this loss.
3. X's action has a high probability of preventing it.
4. X's action would not present significant risks, costs, or burdens to X.
5. The benefit Y can be expected to gain outweighs any harms, costs, or burdens that X is likely to incur.

They think this list helps specify the principle and gives greater direction to the decision.[40]

Beauchamp and Childress establish the obligation of beneficence in terms of reciprocity. It would seem that as such obligations are developed by some form of agreement either from role expectations (e.g., the lifeguard) or from contractual obligations of some form or another. Indeed one might well argue that obligations of beneficence in health care derive from social expectations regarding the profession and from contractual obligations between patients and providers.

Within the context of the principle of beneficence Beauchamp and Childress discuss a number of moral topics from biomedicine and health care. These illustrate some of the difficulties of the middle-level principle approach. For example, they discuss the problems of paternalism within the context of the principle of beneficence even though it can also be discussed under the principle of autonomy. The method, structured around the four principles, will face the difficulty that many moral questions do not fit neatly under one principle or another. Physician-assisted suicide can be discussed and analyzed under the principles of beneficence, nonmaleficence, and autonomy.

In the examination of the principle of justice Beauchamp and Childress are clear that there are many possible specifications of the principle. In fact, at different points they pluralize it by speaking of "principles" of justice.[41] They note that the only common thread to all conceptions of justice is Aristotle's position that "equals must be treated equally and unequals must be treated unequally."[42] This formulation of the principle of justice is formal and in need of more content-full development.

Moreover, Aristotle's formal principle is not the only formal principle of justice. In *The Institutes of Justinian,* for example, justice is defined as the constant and perpetual wish to render everyone his due.[43] But what are persons due? Does justice require them to receive an equal share of resources, opportunities, or freedom? Again, as with other parts of their book, there is no argument as to why we should accept the Aristotelian principle of justice over those of Rawls, Nozick, or any other formulation of justice. It is clear that Beauchamp and Childress do not presume to present a theory of justice. Rather they present a set of assumptions and arguments about justice that articulate a coherent way to adequately address issues of biomedical ethics.

Beauchamp and Childress move to consider alternative material principles which specify the relevant characteristics for determining equal and unequal treatment.[44] The material principles do not have a particular systematic ordering. These material principles are seen by Beauchamp and Childress to be rules of justice. However, the language used can make it seem as though there are actually ten principles, which, like others, specify prima facie obligations that cannot have their weights assessed independent of particular circumstances.[45]

Beauchamp and Childress conclude their method of middle-level principles by treating veracity, privacy, confidentiality, and fidelity as rules justified by reference to the principles and help determine and justify morally required actions.[46] These rules are introduced as specifications of the principles. Specification is an important part of the method insofar as specification is central to how the method addresses particular moral dilemmas. Beauchamp and Childress see obligations of veracity, for example, as being a specification drawn from several of the principles.[47] Thus, they ground veracity in autonomy, fidelity-promise keeping, and the necessity of trust between persons. It is, however, not at all clear from where the obligations of fidelity and trust come. Nor is the relationship of these three rules to one another explained. The introduction of the specifications provides a glimpse of the internal problems of the middle-level principle method. What do the rules mean? How are they related? Why should we adopt this set of rules?

SUMMARY

The appeal to middle-level principles has had a *profound* influence on bioethics. As a field of applied ethics, bioethics has sought to resolve moral controversies by appealing not to foundational moral theories, but to principles common to different moral theories. The principles form the theory of the method. Beauchamp and Childress's four principles have been understood as prima facie binding and equal in weight. The principles are brought to bear on particular moral questions through a process of specification and balancing. The goal of this process is to achieve greater coherence in our moral judgments.

——— The Standard Criticisms ———

The important task now is to assess the strengths and weaknesses of the principlism method of Beauchamp and Childress in bioethics. The principal criticisms of the method in bioethics literature have had to do with three questions. First, how are the principles to be interpreted? Second, how are the principles related to each other? Third, why should this set of principles, and not others, guide our moral judgments? These three questions are the standard criticisms of this approach.

Beauchamp and Childress obviously have captured something important for bioethics in their method. They have successfully articulated, in the four principles, important elements for moral discourse in health care. In this sense the method is very powerful as it articulates many of our intuitions about morality. The principles are content-thin and, in this way they are able to embrace the difficulties of moral pluralism. In this strength, however, lies the weakness of the method. The principles can be conceived so broadly that they take on many different criteria and specifications.

THE MEANING OF THE MIDDLE-LEVEL PRINCIPLES

The first area of criticism has been that the middle-level principles are open to a wide variety of interpretations and, consequently, are of little help in justifying moral judgments. Beauchamp and Childress would argue that the problem is not one of meaning or interpretation but the process of specification. That is, because the principles are content-thin they can be specified in different ways. Some of the critics of the principlism method, notably Gert and Clouser, see the problem as one of meaning and interpretation. That is, the principles are by definition so content-thin that they can take on an enormous range of interpretations. This chapter assumes that the specification and meaning-interpretation dispute is really a different way of labeling the same problem. The crucial question is the sufficiency with which the method handles the problem.

Beauchamp and Childress acknowledge that autonomy, for example, has been used to refer to a diverse set of notions. They try to offset some of the diversity by focusing the middle-level principle of autonomy on actions, not persons. They use the case of *Mohr v. Williams* to illustrate autonomous actions.[48] In the case the physician obtained the consent of Anna Mohr to operate on her right ear, but in the course of the surgery he determined that the left ear was actually in need of surgery and operated on it. A court found that the physician should have obtained the patient's consent, since it is the patient who authorizes the physician to operate to the extent consent is given.[49] Beauchamp and Childress hold that the authorization in the informed consent is an autonomous action of giving permission. However, they have built in criteria (e.g., understanding, absence of control, intentional authorization) that the agent must "substantially" possess to make an act of informed consent. They argue that health care, like other areas of decision making (such as financial planning), should presume a person is competent to make autonomous choices until counter evidence is provided.[50] Under the principle of respect for autonomous choices, they have included a number of complex components, and it is not always clear what the relationship of these components is to one another. It seems as though one is left to one's intuitions and ideas about what they mean in order to interpret or specify the principle. It is not surprising that there is such ambiguity about the meaning of the principle, since there is ambiguity in the term itself.

Respect for autonomy, and autonomous choices, provides a good example of what can be problematic for the principlism method. Respect for autonomy can be interpreted in at least two fundamentally different ways. In explaining the principle of autonomy Beauchamp and Childress cite both Kant and Mill in support of this middle-level principle.[51] For Kant, one acts morally only when one acts autonomously in the sense of willing what reason demands. For Mill, autonomy is shaped by one's liberty and freedom to act. On Mill's account a person is an autonomous agent if one acts in the pursuit of the different desires and goals of his or her life. But such choices are heteronomous choices for Kant, not autonomous moral choices, whenever they are based on any maxim other than a moral obligation. While the same word is used, there are two very different meanings for the term.

Beauchamp and Childress hold the position that one can extract the meaning of autonomy from the different theoretical accounts. They do acknowledge that differences will emerge as the principles are specified. One way to understand the differences that can emerge in the specification process is that the differences are present from the beginning in how the principles are understood. People in different moral communities may use the same term but attach different meanings to the term. Pope John Paul II and Tom Beauchamp may both use the term "freedom" but hold very different understandings.[52] So there should be at least two different place holders for 'autonomy'. There should be 'autonomy$_m$' for Mill and 'autonomy$_k$' for Kant.[53] For it would seem that not every 'autonomous$_m$' action will be 'autonomous$_k$'. They should note the differences, but they go on to say " . . . these two profoundly different philosophers both provide support for the principle of respect for autonomy."[54]

Some have argued that the principle of autonomy can also be treated as a procedural principle; that is, it need not presuppose a particular moral content but only outlines a way in which people determine who has proper authority in a morally pluralistic world.[55] So, the principle of autonomy can be understood either as (1) a procedural principle, (2) a political principle, or (3) a moral principle endorsing a particular moral viewpoint.

This is an example of the coincidence of the strengths and weaknesses of the principlism method. The principle of autonomy is broad enough to serve as a necessary condition for bioethics. At the same time different moral traditions are so rich and diverse that the meanings of autonomy can lead in very different directions. It is also an example of the possibilities of moral acquaintanceship. There are arguments and disagreements surrounding the concept of autonomy. The methodological question is whether or not it is best to approach autonomy by way of a principle. Because of the problem of interpretation of the principle, I will argue it is more helpful to approach issues of autonomy through procedures of consent.

The problem of interpretation is also seen in the other principles. Nonmaleficence, for example, has a wide variety of meanings. The problem of clarifying its meaning and specification is not resolved by appealing to concepts of "harm" or "interest"[56] as these terms have many meanings. Beauchamp and Childress seem to offer an interpretation of harm as tied to basic welfare interests. However, the principle, as such, does not specify this as its only meaning. Recall that Peter Singer defined "interest" as suffering or experiencing enjoyment or happiness. Moral languages may often use the same words but with very different interpretations. This is why the principle of autonomy, or the other principles, will have a wide array of interpretations and specifications.

The principle of beneficence must confront the same problems of interpretation and specification. Beauchamp and Childress note that beneficence can mean the provision of benefits and/or the achievement of balancing benefits and harms for patients. They choose the second, contending that the principle of beneficence grounds the practice of biomedicine.[57] However, they rely on a notion or vision of the 'good' and its implications for patients that is not specified.

A central problem with the principle of beneficence in the principlism method is the difficulty in determining the limits of beneficence. This is a particular problem for the authors or anyone who wishes to treat beneficence as an obligation and not as an act of charity.[58] This problem is indicative of the difficulties with the middle-level principle approach when it is used in a dilemma where the participants have diverse views of the meaning of the principle.

In examining the example of the drowning man one can ask: "Why am I *obligated* to 'do something'? Why isn't my attempt to rescue appropriately described as an act of charity or a supererogatory act of moral heroism?" These questions make clear that with a free-standing principle, having no definition of its scope and strength, there is no reason to conclude that such an obligation exists. There seems then to be no clear statement of the principle. The difficulty is that there is no clear boundary where obligation ends and supererogation or charity begins. However, the attempt to set limits to the principle indicates that the limitations and content of the principle are derived more from social and professional roles than from the principle itself.

The roles themselves are specifications of the principles. Here the Pellegrino and Thomasma account is an instructive foil. They address beneficence in the practice of medicine. That is, the social context of the practice of medicine defines the concept of beneficence.

The problem of interpretation, or specification, is most clearly seen with the principle of justice. Each interpretation of the principle of justice is the root of different ethical theories.[59] Each highlights a material principle that is important for moral reflection. As with a principle of autonomy, a principle of justice can be interpreted in a number of diverse ways. In contemporary moral and political philosophy there are very divergent views of justice and its meanings. John Rawls offers a content-full theory of justice while H. T. Engelhardt, Jr., offers a procedural account of it that he thinks is content-thin. In these different views the problem of meaning and specification becomes clear. To the extent that there is diversity about the meaning of the principle there will be significant differences in interpretation and specification. Just health care for Rawls will look something like Norman Daniels's view of a basic system, which is available to all and offers fair opportunity to all. Just health care for Engelhardt will be more like a free-market system. Beauchamp and Childress acknowledge the different approaches to justice and opt for a Rawlsian account of fair opportunity in health care. They think the principlism method is the best reconstruction of common morality. While principles capture the broad parameters of morality, it is hard to see how they are helpful when they can be specified in so many diverse ways. One might worry that the method will add to the confusion rather than help to clarify situations.

One possible way to come to terms with the interpretation of moral principles is to see them in relationship to the rules they generate. The difficulty of interpretation emerges for principlists in that there will be conflicts between the principles and it is not clear that the use of moral rules can resolve these conflicts. Moral rules are, generally, regarded as more specific action guides than the principles from which they derive.[60] It seems that rules specify types of actions that fall under particular principles. For example, the principle of respect for autonomy may give rise to several rules (e.g., "It is wrong to lie" or "It is wrong to use force against the innocent"). For Beauchamp and Childress the rules are part of the processes of specification and balancing. The rules help to specify principles to particular situations while the appropriateness of the specification is judged in terms of the process of balancing and achieving coherence with the other principles and rules. The problem of meaning is often transformed at the level of rules. One might see a particular rule, for example, tied to more than one principle. Rules about truth telling could be tied to autonomy and justice, for example. So the problem of interpretation is not always solved by specification. Rather, it can be shifted to the discussion of rules. We are left with the problem of identifying the rules that belong to the principle, and we are ignorant of the relationship between the rules and how they are to be employed when they conflict.

In determining the interpretation of a principle Beauchamp and Childress often illustrate the middle-level principle through cases and moral problems. The cases are used as a way to illustrate the application and interpretation of the principles. For example, in their discussion of the principle of nonmaleficence Beauchamp and Childress include a discussion of the treatment of disabled newborns in which they say that when there is doubt, one ought to err on the side of preserving life.[61] This interpretation of the principle tells us to whom the principle applies (the newborn). They appeal to a reasonable-person standard as the basis for how treatment decisions should be made for patients who were never competent, such as disabled newborns. But as one looks at the case neither the principle nor the case spell out that "not doing harm" means saving a life in such cases. Perhaps the principle means forgoing treatment to allow relief from long-term pain and suffering for the newborn. Further, they give no reason why the scope of the application of the principle in this case is only to the newborn. Such cases can cause harm to families in terms of costs, burdens, and suffering and harm to society as well. In the development of the principlism methodology Beauchamp and Childress have adopted, more explicitly, a use of Rawls's process of reflective equilibrium. In this process the cases influence the norms and how they are specified.

In summary, a strength and weakness of the principlism method is that the principles are content-thin. They are thin enough to be general in appeal. However, the principles need content in order to resolve the issues of bioethics. Beauchamp and Childress use the processes of balancing, weighing, specification, and coherence to link the principles to the issues in bioethics. However, each of these elements carries moral commitments. This is why many critics have argued that the meaning, interpretation, and relationship of the principles are points of controversy. By incorporating the elements of specification, balancing, weighing, and coherence Beauchamp and Childress bring in other methods such as virtues, cares, and rules. Indeed, since abstraction, which is at the heart of formulating the principles, is viewed as a process of generalization, it is not surprising that the principles should be analogous to chapter headings encompassing a wide variety of meanings.

THE RELATIONSHIP OF PRINCIPLES

Another standard criticism of the principlism method is that it does not articulate the relationship of the principles to one another. A range of interpretations for the principles sets up a range of possible relationships. An obvious question is how they are to be applied when they conflict. One can easily imagine cases in which the principles of beneficence and autonomy conflict with one another. "Dax's case" provides just such an example where the patient's choices, at the time of the intervention and following, were in conflict with what the health care providers saw as their duty as defined by the patient's best interests. In a discussion of the case, Childress has argued that Dax's autonomy at some point takes precedence over the beneficent actions of others. But in the middle-level principle approach there seems to be no *reason* for one principle to take precedence over another.[62] The choice to treat, justified under the principle of beneficence, conflicts with a choice that follows the patient's wishes, justified under the principle of autonomy. It can even be argued that Dax can appeal to the principle of nonmaleficence. That is, he has been harmed by being kept alive in a way that he finds burdensome. The approach of the principlism method relies on balancing and specifying. Given that the principles are content-thin (i.e., they do not have a particular content or moral point of view) and absent any particular ranking, the mechanisms of specification and balancing are crucial.

One strategy for resolving conflicts of principles is to have some type of lexical ordering of the principles (e.g., Veatch, Rawls). Beauchamp and Childress, however, eschew such an ordering. Without an ordering the principlism method relies on some form of intuitive balancing of the principles. The weight given to a principle, in relationship to others, seems to be assigned according to the case. Beauchamp and Childress assert that each principle has 'weight' but they do not assign a priority weighting or ranking.[63] However, it seems odd to use the language of 'weighing' without developing a common measure by which weights are assigned and compared.

Such a measure is never developed in this method of principlism. It would seem that all the principles are equal in moral decision making. In cases of a conflict, consequences, rights, and justice all compete equally. In their brief discussion Beauchamp and Childress say that the relative merit of each principle will be determined by the particular case.

This involves a covert appeal to intuitions and moral judgment and no full account of moral judgment is developed in their work. While their approach might be similar to Ross's intuitionism, they believe they go beyond Ross in offering a procedure of moral reasoning that reduces, but does not eliminate, intuition.[64] As the principlism method has developed, it follows a model similar to that of Rawls. The model begins with our considered judgments and strives toward an ideal of coherence. Coherence is the target for the specification and balancing of the principles. The Beauchamp and Childress account does not provide us with a basis for knowing how to employ the principles in actual cases where more than one principle may be relevant. However, it is worth noting at this point that the principlism methodology can be very helpful in the area of policy development. This is an important point taken up in the second part of this book.

Their discussion of paternalism provides a glimpse of the problems inherent within their approach. In their discussion of the meaning of 'paternalism', the authors adopt, for reasons not given, a definition from the *Oxford English Dictionary* in which the father (professional) acts beneficently in accord with his conception of the interests of his children (the patient) and makes decisions for them.[65] Beauchamp and Childress distinguish types of paternalism (strong and weak) and conclude that while strong paternalism is usually wrong, it may nonetheless sometimes be justified and that weak paternalism is justified when the person is not autonomous and is in danger. Paternalism brings the two principles of beneficence and autonomy into direct conflict. Beauchamp and Childress appeal to an equal weighting of the principles and conclude that "[i]t therefore cannot be assumed that even an autonomous choice can never be validly overridden on the grounds of beneficence to the chooser. . . . "[66]

This difficulty of conflict between the principles may not be obvious because of the adroit way in which Beauchamp and Childress employ cases. They emphasize the role and place of particular cases in

reaching and justifying moral judgments. This is an important methodological ambiguity in their project since they neglect to articulate
an understanding of cases and the role they play in moral judgment.
They provide cases which appear appropriate and which seem to illustrate the use of the principles. Yet how ought one select relevant
cases? How are cases to be 'read'? In their discussion of cases and casuistry[67] Beauchamp and Childress point out the importance of an interpretative structure (the principles) for understanding cases. (This is
a point in which I am in substantial agreement with Beauchamp and
Childress. See the next chapter.) However, the interpretation of the
principles returns as an important problem in moral judgment. How
the principles are understood and interpreted will influence the description of the case. Often the very crucial problem in bioethics is
the lack of agreement about a description of a morally controversial
case. I will later argue that the very description of cases, along with
their selection and characterization, is a complex moral task influenced
by our moral views and ranking of moral principles.

FOUNDATIONAL QUESTIONS

Finally one may ask: "Why these principles?"[68] Beauchamp and
Childress think that the four principles are the basic articulation for
common morality underlying biomedical ethics. However, there are
other principles that are often cited and used in the field of biomedical ethics. Indeed, if one looks at other works in biomedical literature
one finds other lists of principles. Some, for example, might cite the
principle of the sanctity of life or a principle of patient advocacy as
crucial principles for biomedical ethics. Other principlists, like Frankena
or Veatch give different accounts of the principles and their relationship. Some may argue that the differences between them are not significant and that one could simply collapse the list of principles in
Beauchamp and Childress into Frankena's list. However, Beauchamp
and Childress do not share this view.[69] It is also important to note that
Frankena understands principles to be grounded in the moral point
of view and the principle of "love," while Beauchamp and Childress
understand their list to be grounded in ordinary moral discourse.

It is fair to ask, however, whether or not there is a common moral
discourse in bioethics, or whether there are numerous moral dis-
courses, or whether there is moral "babel," or something between the

extremes of babel and coherence. The second part of this book argues for something in between the two. That is, I will argue that there are overlapping moral commitments that allow discourse and argument between *moral acquaintances.* It is the overlapping yet different aspects of moral acquaintanceship that allow for both the possibility of discourse and the criticisms of differing positions.

In the most recent edition of their book Beauchamp and Childress have expanded their account of the method of principlism and the foundation of its content in common morality. The account of common morality they deploy is content-thin and seeks to incorporate moral pluralism and diversity. Thus they can hold for a common morality and allow for the deep divisions in secular society on fundamental moral issues such as abortion, physician-assisted suicide, and embryo and fetal tissue research.

One problem with their method is whether or not the common morality is so thin that the concept is useless. Put another way, are the principles so general in meaning that specification is bound to produce conflict? There are, of course, crucial questions about the appeal to common morality that cannot be settled. First, there is the question of whether or not there is a common morality. Certain thinkers, like Alasdair MacIntyre, Jean-François Lyotard, and H. Tristram Engelhardt, Jr., argue that there is no common morality for contemporary society. MacIntyre gives insight as to how to understand the lack of common morality in contemporary Western societies. In *After Virtue* he speaks of men and women sharing the bits and shards of a broken past which they are unable to reassemble. They are shaped by a past that they do not understand. In *Whose Justice? Which Rationality?* MacIntyre speaks of those men and women who are simply rootless cosmopolitans with ties to no particular moral community past or present. Finally, one can also find men and women who understand themselves within the context of intact moral communities and traditions. This diagnosis of contemporary moral discourse enables us to understand why people look at contemporary society and think that either there is a common morality or there are only moral differences.

This alternative account, where there are only bits and shards, does not adequately address why we have been able to achieve agreement on certain issues and why there can still be moral debate. We need another account and it will be one that is not at all clear cut. There may be something like a common morality but it is not easily captured.

It resists general formulation in any particular way. This may explain why Beauchamp and Childress think it necessary to include accounts of casuistry, virtue theory, and the ethics of care along with their account of principles. If one examines the development of the different editions of *Principles* one can find how the method has become more and more inclusive of other methods. The questions about common morality and the phenomenon of agreement need much more examination, and they are the focus of chapters 4 and 5. Yet, at this point, in assessing the middle-level principles, it is essential to highlight the assumption of a common morality.

In the development of their model of middle-level principles Beauchamp and Childress, in part, recognize the historical context of the principles.[70] In the most recent presentation of the model they endorse a "robust historicism" of moral standards and principles. While they recognize the communal, cultural, and historical contexts of moral standards, they nonetheless hope for an overlap of principles and norms. The assertion of a common morality raises a second question. Even if there is a common morality, why should we assume that it is morally normative? Traditionally we have sought to answer the question with some foundational view that provides a critical point. However, I have been arguing that such a position, outside of our normal moral practices, does not exist. Rather, we need to begin where we are in "morality." Critical questions can be raised by others.

SUMMARY

The methodological appeal to principles stakes out basic, necessary conditions for morality for bioethics. The principles are content-thin, which allows them to accommodate a wide spectrum of meanings and interpretations. Beauchamp and Childress argue that a morality without these principles is incomplete. In addition to completeness, the method also relies on the specification of the principles to particular moral situations and questions. The method also relies on balancing and coherence.

Critics have raised questions about the meaning of the principles, their relationship to one another, and whether or not the principles are the best articulation of morality. Beauchamp and Childress can respond by arguing that their method captures the basic conditions for morality and that one cannot look for more certainty and specificity than the subject matter will allow. Moral pluralism, in this method, will be incorporated in the wide range of meanings given to the principles. In the method there can be argument and discussion to the extent that men and women share similar moral commitments. The less that is shared the more the argument will center around methodological considerations of coherence, specification, and balancing.

—— POSTMODERNITY AND PRINCIPLES ——

Postmodernism, as used here, refers to the end of the philosophical search, associated with the Enlightenment, for a common, normative, content-full morality. It is the recognition that there is, in Beauchamp and Childress's language, a common morality with many interpretations. Later chapters will argue that "common morality" may be misleading in that some will assume common morality to be far more robust than others. It is more helpful to start from our common procedures to articulate a framework of common morality for moral acquaintances.

COMMON MORALITY AND THE MORAL POINT OF VIEW

A crucial assumption of the middle-level principles project is the existence of a common morality. The principles are understood as embodying the basic features of common morality. For Beauchamp and Childress morality "refers to social conventions about right and wrong human conduct."[71] These conventions are widely shared such that they form a stable communal consensus. The implicit argument of the project is that the middle-level principles capture the necessary conditions for common morality and, in so doing, are able to address moral dilemmas while offsetting the disputes about the foundations of moral theory.

This claim about a common morality can raise more questions than it answers. One question is whether or not a common morality exists. This is an empirical question, and the answer will depend on the evidence one finds. Issues like abortion, population control, embryo research, assisted suicide, along with questions of public policy about justice and health care resources all point to less commonality than Beauchamp and Childress need to assume for their project to work. They would argue that the account of common morality they deploy is content-thin and accounts for moral pluralism. Thus, they would continue, no one should be surprised at the deep differences that exist on such issues. Then, however, one has to wonder how helpful the appeal to common morality actually is.

But differences are only part of the picture in bioethics since the history of bioethics in the last thirty years has witnessed a number of examples of resolution of moral issues. Many issues in research, experimentation, and the place of informed consent in the clinic are examples of resolutions. In the second part of this book it is argued that the question of common morality is not a simple one to answer. Common morality is lived out in many different ways by different moral communities. This is evidenced by moral pluralism in secular societies. By understanding the role of the moral community one can better understand why there are great moral differences in secular societies as well as points in common. There are points in common that enable people to understand one another, reason about issues, and, at times, reach agreement. These are the achievements of moral acquaintances. To the extent that there is acquaintanceship, it is possible for principles to work. However, there are many areas where there is little in common. In those situations the principles will be much less helpful.

Another crucial question is whether or not common morality is best captured by the middle-level principles. Beauchamp and Childress acknowledge that there are other elements of this morality as they seek to include other moral appeals such as the virtues and the ethics of care. One objection would be that in an effort to claim a common morality Beauchamp and Childress have suppressed differences in a search for sameness. These different appeals actually represent very different moral views that have often been suppressed. That is, the idea of a moral principle is not always self-evident. How principles, qua principles, are understood reflects an understanding of the moral world.

WHAT ARE PRINCIPLES?

One way to understand and assess the principlism method in bioethics is to explore what it is to be a principle. That is, how have principles been deployed in the history of ethics? Such an account, while not definitive, can be instructive.

'Principle' comes from the Latin *principium* ("take first") and its primary meaning is that of "beginning," "commencement," and "fountainhead." A principle is thought to be a foundational source from which thought proceeds. The term 'first principle' is employed in the Western philosophical tradition in a fairly consistent way. Aristotle speaks of first principle(s) as the first point from which something is or comes to be known.[72] Saint Thomas Aquinas followed this and distinguished principles of knowledge and principles of being. In his systematic reflection on ethics Thomas distinguished between primary principles and secondary principles.

Primary principles were understood as a framework for the moral sciences and so fundamental that there was little room for dispute.[73] Thomas characterizes the precepts of the natural law as analogous to the first principles of speculative reason; that is, they are both self-evident. The first indemonstrable principle of reason is that "the same thing cannot be affirmed and denied at the same time, which is based on the notion of being and not-being: on this all others are based."[74] As 'being' is the first principle of speculative reason, "so good is the first thing that falls under the apprehension of the practical reason, which is directed to action."[75] The first principle of practical reason is that "good is that which all things seek after," which leads to the first precept of the natural law that "good is to be done and pursued, and evil is to be avoided."[76] Thomas goes on to set out other primary precepts: preservation of life, inclination to the truth, and faculties and powers are to be used according to their natural function.

Secondary principles are particular conclusions concerning types of actions drawn from the primary principles. An example is "[a]dultery is forbidden." While the primary principles are the same for all, the application of the secondary principles must consider the details of particular cases and may be changed in some circumstances.[77]

If one develops an Aristotelian-Thomistic approach to principles, the strengths and weaknesses of the middle-level principles become easier to handle. This is because the Aristotelian-Thomistic approach provides a theoretical account along with a ranking of principles. Certain principles have a preeminent status, which informs and shapes the secondary principles. This ranking gives a direction for specifying and balancing the principles. The meaning of the secondary principles is illuminated by their relationship to the primary principles.

One could argue that the method of principlism is an attempt to form a theory from the secondary principles. The principles will have applicability insofar as those who use them share enough in common

to specify and balance them. The middle-level principle approach eschews grounding in some first principle(s). This forces the middle-
level principles to do the work of the first principles; that is, to ground
the rules that follow. To the extent that this can be done successfully,
as Beauchamp and Childress think it can, bioethics can have moral
theory without foundations.

The scholastic notion of primary and secondary principles is not
the only model by which to understand the role of principles in moral
justification. This historical example just illustrates another model of
principles, and it suggests one way to understand some of the criticism of the middle-level principles. The extent to which they succeed
or fail will depend on the extent of a shared moral context.

One discovers similar considerations advanced in other approaches.
For example, in utilitarianism the principle of utility fulfills the function of a single first principle, which structures the rules or judgments
about acts. This first principle allows one to understand that the central concern is the maximization of utility, however defined, and allows an understanding of the reason for rules and their relationship to
one another. For example, a utilitarian may have a rule about telling
the truth which directs her to maximize utility for the individual and
society. Kant's first principle is the categorical imperative. It establishes
the theoretical machinery for deriving secondary principles and rules.

Historically first principles have been employed in grounding moral
theory with precepts and rules deduced from the first principles.
Beauchamp and Childress, relying on a particular notion of "abstraction," employ the secondary or middle-level principles as generalizations and summaries of a variety of disparate moral insights. There are
two senses of 'abstraction': (1) identifying the 'essence' of something,
or (2) creating a generalization or summary. Beauchamp and Childress
opt for the sense of generalization, and their principles become like
summaries that attempt to bring together under one label a number
of insights that do not cohere one with another. The process of specification, balancing, and weighing are used to minimize incoherence.
It is the process of specification and balancing that give the principles
coherence rather than the foundations of the theory.

Still, Beauchamp and Childress might appeal to an earlier historical
example as a precedent for the role of middle-level principles. In the
first four centuries of Western Christianity the Christian community
was without the complex theological structures that emerged during

the High Middle Ages. Moral life was structured by a set of rules (e.g., the Decalogue and ecclesiastical canons) applied to the particular cases brought by individual penitents. This seems to be similar to the type of structure Beauchamp and Childress presuppose for the principles of biomedical ethics. There are, however, two significant differences. First, the moral rules of early Christianity had a clear and firm *foundation* in the Tradition. There was a concrete understanding about which set of rules was to be used. The rule was also set within the context of the life of a community and its exemplars. Second, the problems of interpretation were eliminated by the local bishop, or his delegate, who held the authority to interpret the rules. So there was a clear mechanism for interpretation. Finally, there was a mechanism for application in that the bishop set the guidelines for confessional practice and moral advice for the local confessors. In Beauchamp and Childress we have rules but the context of faith, the common morality, within which to understand them is ambiguous.

This example from early Christianity is instructive. It provides a picture of an actual moral community's use of principles and cases. One understands the role of the principles because there is a defined moral community in real circumstances with real issues, which provide a flesh and blood content to its moral beliefs. The secular community Beauchamp and Childress address is ill-defined, pluralistic, and contentious regarding its moral beliefs. This makes the interpretation, specification, and balancing of the principles, all essential to the method, somewhat tenuous. No method, save an absolutist one, will escape some degree of tenuousness. However, in later chapters I will argue some of the ambiguity can be alleviated by analyzing moral communities, the basis of lived moral experience, and the common procedures of a secular society.

PRINCIPLES, CONTENT, AND AMBIGUITY

In some of the literature concerning middle-level principles one finds attempts to articulate an understanding of moral principles. Childress, for example, notes three characteristics of 'principles': prescriptivity, universalizability, and supremacy. R. M. Hare focuses on similar criteria.[78] But these characteristics, by themselves, will neither disclose the moral content of middle-level principles nor help us understand how the principles should work. The characteristics will not distinguish moral principles from those of etiquette. Childress also identifies a fourth criterion for "material content, particularly one that relates what are moral and nonmoral principles."[79] The fourth criterion is essential to understanding moral principles qua moral principles. Childress calls it "otherregardingness."[80] Even if there is agreement about this criterion there would still be the question: Who is the other? Singer, for example, would want to include animals, and not fetuses, as "others." Those who are Pro-Life will want fetuses included. The point is that even the most basic assumptions about the "moral" world assume a particular content. One could just as easily argue that the moral point of view is self-regarding.

A utilitarian, in contrast, will define the moral point of view in terms of some form of utility (e.g., achieving the greatest good for the greatest number or satisfying the greatest number of preferences) and moral rules will have a prima facie standing. They will be interpreted and applied in light of the definition of utility.

Without a shared interpretative matrix it will be impossible for someone to know how the principles are understood by others. Each person will understand a moral principle within her / his own interpretative matrix. While using the same moral terms men and women can mean very different things. They may use the same terms but speak different languages. It is possible that in the use of moral terms men and women can be like two scientists from two different eras discussing 'mass'. One is a Newtonian where 'mass' is absolute in space and time while the other is an Einsteinian for whom 'mass' is an alterable part of a relative system of space and time. Though they use the same words, their meanings are incommensurable. In this way the middle-level principles function as place holders in heterogeneous languages. This does not mean, however, that the principles are unintelligible to people with different moral views. As argued in the second part of this book, there are overlapping meanings and values that allow men and women to be moral acquaintances.

The same language problem emerges in relating the principles to cases. Moral cases must be set in a language if they are to be identified and resolved. For example, the prohibition against rape can be described in two different ways. The utilitarian prohibits it because it creates disutilities in the world by going against the preferences of the victims and those who would sympathize with them more than it satisfies the preferences of the rapists and those who would sympathize with them. But since no act is intrinsically wrong for the utilitarian, neither is rape. In contrast the deontologist may argue against rape because it fails to regard, properly, the victims and their autonomy. The act has intrinsic wrong-making characteristics. It is wrong regardless of the consequences. Thus, although we have agreement on a prohibition, the descriptions of the case and the reasons for the prohibition are quite different. The accounts are given in different moral worlds. The descriptions and justification of the prohibition against rape are developed from two different phenomenologies of moral experience. So too with suicide. A utilitarian might argue that suicide ought to be prohibited because of the disutility it produces for the society as a whole while a deontologist would argue that it ought to be prohibited because it is wrong in itself or permitted because it cannot be known to be morally unobjectionable in itself. The utilitarian prohibition is open to change and revision while the deontological one is not. The two prohibitions are not the same. They can only be understood in the context of the moral worlds from which they come. Definitions do not exist apart from the web of language.

———— Principles, Cases, and the Moral Point of View ————

The appeal to middle-level principles is conceived as a way to develop an understanding of biomedical ethics in the face of the failure of the foundational project of grounding in a general, secular morality. This appeal has been criticized because of difficulties in understanding the meaning of the principles invoked, their relationship to one another, and their application to cases. The principlism method has met with success and failure. Both can be undersold through an examination of the role of moral community in shaping content-full views of morality. Insofar as there are overlapping values and moral commitments the principles may work. To the extent that there are differences the principles will not resolve moral dilemmas or controversies. Roman Catholics, Orthodox Catholics, and Jews can reach a common position on the morality of abortion. However, the same set of communities will disagree about the use of in vitro fertilization.

One can well appreciate these problems in terms of examples from the use of 'principles' in the history of moral philosophy. Moral philosophy has often had recourse at the foundational level to first principle(s) of some sort (e.g., scholasticism, utilitarianism, deontology). Within intact moral traditions one finds principlism embedded in a framework. In contrast, the middle-level principles method attempts to move on the secondary level without the orientation and support of first principles or a broader common framework. Support for the method depends on common morality. When common morality is thin or absent the meaning, relationship, and application of principles can no longer be specified or balanced. The extent to which the method can work depends on the context of common morality.

What Rawls once wrote of rules applies to principles.[81] He argued that when one plays a game one understands the rules as part of the game. Outside of the game the rules do not have the same meaning as they do within the game. In a similar line of reflection Wittgenstein showed that rules do not interpret themselves. They are part of a way of life and need to be understood within that context. Like rules, principles are bounded in customs and social institutions and contexts.[82] In his later writing, Rawls considers the social and political context of the principles of justice.[83] To confront the problems of interpretation one must situate moral principles within the context of a sphere of meaning in which the rules or principles are developed. Beauchamp and Childress recognize the historical, communal context of moral principles. Many of the most controversial issues in bioethics—nutrition and hydration, genetics, abortion, fetal tissue, access to health care—raise an important question of whether or not there is a common context for these issues.

Beauchamp and Childress, in trying to elucidate the meaning of the principles, appeal to cases as the context in which principles are weighed against one another. This appeal seems incomplete but it points us in another direction for understanding and justifying moral judgments. Some would argue that the common morality Beauchamp and Childress attempt to capture in the principlism method is better articulated by a turn toward cases. Recalling Rawls's game analogy, within the context of a game certain situations (cases) are held up as paradigmatic. A theory may provide the formal rules of a game (e.g., three strikes make an 'out' and three 'outs' to a side per inning). But there are also 'informal' rules of a game (e.g., outfielders should move in on a 'bunt') that are just as important to the game as the formal rules. When learning a game one not only learns the formal (procedural) rules but one learns strategies as well (the ways to respond when certain situations present themselves). The boundary conditions are set by the formal rules, which form the broad outlines of what situations are possible. They set out the procedures for how the game is played. Batters, for example, cannot take a base without a hit, a walk, or being hit by a pitch. The judgment of players, as to how to respond to a situation, is shaped by the formal rules and the informal strategies. Their judgments will depend, in part, on the content they received from their coach.

By analogy, moral reasoning does not only appeal to principles and formal rules. It often appeals to paradigmatic cases (i.e., in certain situations one should do x) whose descriptions are shaped by the moral rules and principles one follows. Indeed, one of the criticisms to the use of middle-level principles in bioethics has been that the approach fails to appreciate the case-oriented nature of bioethics and moral reason. Some have argued that bioethics should begin with cases in order to articulate principles.

This method of casuistry needs to be explored. However, before turning to casuistry it is important to note, and not forget, the more fundamental challenge. The challenge of casuistry, from the point of view of the casuists, to the method of principlism concerns the very nature of moral rationality. This is an issue raised by many writing in feminist ethics as well. The most foundational challenge to the principlism method is not the pluralism of moral content but the pluralism of views about moral rationality. The two are bound together.

4. After Paradigms:
The Crisis of Secular Casuistry

Bioethical cases can be like mercenaries. They seem to have no loyalties to a cause, nation, race, or principle. They can be used by anyone. Cases often appear in different methodologies. However, it is not always clear what role cases of moral controversy actually play in different methods of bioethics. In much of the literature of bioethics cases function as expository devices; that is, they function to illustrate a theory or principle. At times they seem to be used almost as an afterthought to illustrate theoretical points. Different methods in bioethics use cases in different ways.

It should not be surprising that cases should be so attractive for methods in bioethics for it is a discipline that has arisen in response to the moral controversies in medical practice and the framing of health care policy. The cases of Quinlan, Cruzan, Jobes, Conroy, Quill, and Dax are essential literature in bioethics. The field has played the role, in public courts, in the clinic, and in hospital ethics committees, in attempting to resolve cases. Moreover, the practice of medicine is itself case driven. This concern with concrete cases has led, in recent years, to the development of a literature on casuistry in bioethics: a case driven methodology for resolving moral controversies. Casuistry assumes cases to be more than merely illustrative of moral theory. The method assumes that cases are the source of the methodology. In the contemporary world, with many different moral theories, casuistry's pragmatic focus holds a natural appeal for resolving the dilemmas of bioethics since it offers a hope of resolving actual cases without recourse to theory or principles.

In reflecting on the notion of a "case" one can begin to outline the difficulties general, secular casuistry will encounter in practice. The root of case is found in the Latin verb "cade" meaning "to fall." The term conveyed a sense that a case was an instance of a class of things or events insofar as a case fell under a general description. Moral cases are instances that fall under some moral description (e.g., "This is a case of theft"). For casuistry the resolution of moral dilemmas depends on linking the case at hand with an appropriate moral description. However, the moral controversies of contemporary, secular bioethics are not easily resolved by casuistry because the level of secular moral discourse often lacks a common way to describe such cases and reach some resolution. In secular, morally pluralistic societies there are many ways to describe cases and many ways to rank them in relationship to each other.

This chapter will explore the implications of moral pluralism for casuistry and the possibilities of casuistry as a method for bioethics. Too often those attracted to casuistry's pragmatism have overlooked the pluralism of casuistries found in both its history and the contemporary secular world. Moral intuitions and values shape the characterization of cases, influence the very selection of cases that need resolution, and identify the cases that are to be paradigm model cases that guide and direct moral choices. As moral virtues are realized within the context of the particular cultures or set of moral values, so too a casuistical method needs some moral sense or moral authority to identify and resolve cases of moral controversy. The proponents and practitioners of casuistry often overlook the moral commitments necessary for the practice of casuistry. However, in the framework of a morally pluralistic society such assumptions become much clearer and more open to discussion. There will be many casuistries.

Diverse models of casuistry have become prominent in the literature of bioethics.[1] Jonsen and Toulmin have a different approach to the practice of secular casuistry from Baruch Brody. Moreover, case reasoning is not confined to bioethics. In general ethics there are prominent examples of casuistry that focus on the resolution of particular cases.[2] These analyses, however, do not attempt to develop a general theory of casuistry. To understand the difficulties, possibilities, and limits of secular casuistry one must move from cases and think about the necessary conditions for secular casuistry.

The recent renaissance of casuistry and different forms of case reasoning is another response to the limits of foundational methods of moral reasoning to satisfactorily resolve controversies of moral reasoning. In part the case methods are a response to the general, abstract nature of foundational methods and principlism. Bioethics has been a field marked by controversies that are often intractable. One has only to call to mind the interminable disputes and disagreements that arise around issues such as abortion, assisted suicide, or embryo research. In such areas of bioethical controversy the application of moral theory has not achieved an intellectually authoritative resolution for many. Even appealing to middle-level moral principles does not resolve such controversies. For example, in questions surrounding compulsory HIV or drug testing there are often conflicting appeals to principles of justice, beneficence, and autonomy. Many have argued that a "top-down" method of moral reasoning, in theory or cases, is inappropriate for ethics. Rather, what is required for moral thought is some kind of "bottom-up" model that begins with the particulars of a moral question.

Jonsen and Toulmin argue that the often interminable discussions in bioethics are due to a misunderstanding of moral reason. They argue that moral reason should not be understood as theoretical reason but as rhetorical, practical reason. It is in the discussion of cases that moral controversies are best resolved. Jonsen and Toulmin illustrate their view by examining the ancient roots of casuistry and its development in Roman Catholic moral theology. Jonsen and Toulmin have made an important contribution to bioethics and the discussion of methodology by identifying a different perspective for initiating a method. The method focuses on a view of moral rationality and justification that is different from other starting points and assumptions in bioethics. Implicit in the Jonsen and Toulmin criticisms of moral thought are questions for their own approach. They raise questions about which account of moral rationality ought to be used in bioethics. However, it becomes clear that there are different models of case-based reasoning and it is not clear which model of case reasoning should be used. Furthermore, within their model of casuistry there are important questions. The strength of the method—its use of cases—is also its weakness. In a morally pluralistic society the selection of paradigm cases will be problematic for casuistry. For example, is assisted suicide a case of care and compassion or a case of murder?

In examining the Jonsen and Toulmin account I will argue that their hope to revive this model is far too optimistic. In part, their effort fails because they do not adequately account for the structures of moral authority which were and are essential to the model they choose to sustain their argument. Second, the history of casuistry they present is incomplete. The model of casuistry they have chosen is highly structured both in terms of its moral commitments that give it content and the sociological context that guides case by case decisions. Third, the model of casuistry they deploy relies on paradigm cases and common structures of moral authority for case interpretation (e.g., confessors, bishops) and therefore cannot succeed in a secular context that is morally pluralistic, directed by different moral views, and governed by no canonical structure of moral authority beyond that of individuals. If one thinks of morality as part of a way of life then one must attend to the sociology of moral knowledge that shapes particular cases. In a secular society that celebrates its cultural pluralism, there will be differing sociologies of knowledge and understandings of cases.

It may be argued that these criticisms of Jonsen and Toulmin's method cannot be extended to other models of case reasoning. In part this is true, for the method of casuistry selected by Jonsen and Toulmin is highly particular in both content and method. However, in critically examining their approach, we can gain important insights about the conditions needed for a successful use of casuistry. To illustrate the conditions for casuistry I tie my critique of Jonsen and Toulmin to the discussion of the "Trolley Paradox."

Again and again one returns to the crucial point that casuistry can only be successfully practiced if there is a common moral sense or set of common intuitions or structures of moral authority in order to identify exemplar cases to enable discussants to share a sense of a controversy and resolve it. A secular, morally pluralistic world will have diverse moral cultures and no common moral authority. When a secular casuistry is successful, it is because of a coincidence of moral commitments. This overlap of views of cases is important and needs to be examined. The phenomenon of overlapping moral consensus is explored in the category of moral acquaintanceship developed in the second part of this book. This chapter will not only set out the limits of secular casuistry but its possibilities as well. These possibilities are better understood in the framework of moral acquaintanceship (chapters 5, 6, and 7).

—— JONSEN AND TOULMIN AND ——
THE RENAISSANCE OF CASUISTRY

By their own account Jonsen and Toulmin became casuists from working for the National Commission for the Protection of Human Subjects of Biomedical and Behavioral Research (U.S.A.). Rather than proceeding "from the top" by applying ethical theory to practical matters, the commission worked "from the bottom" articulating moral maxims and rules in particular cases. Only later, after finishing its work, did the commission construct a theoretical framework for its work when it authored the *Belmont Report.*[3] Jonsen and Toulmin see the deliberation of the National Commission as an example of a casuistical methodology. It is, for them, an example of the successful resolution of a moral controversy by casuistry. They hold that such a method is more closely identified with how we actually think our way through moral quandaries and reach judgments. However, these claims about the work of the National Commission need to be carefully examined. Beauchamp and Childress interpret the work of the same commission as an example of the principlism method. It is clear that the work of the commission can be accounted for in fundamentally different ways.

Fortified by their experience, Jonsen and Toulmin set out to develop a model of secular casuistry. Their book has at least two goals. The first is to identify the interplay of principles and cases within the process of description and re-description of moral quandaries in order to resolve moral dilemmas. For bioethics and other areas of applied ethics this goal is not controversial. The second, and more interesting, goal of the book involves a much stronger claim for casuistry: establish the primacy of moral cases over moral theory or principles. They further argue that casuistry, as seen in the work of the National Commission, can achieve agreement across theoretical divides—a feat in their view that neither the methodology of theory nor middle-level principles has been able to achieve. They are claiming thus to be able to resolve the problems of providing moral foundations for bioethics.

THE NEED

Jonsen and Toulmin argue from a particular model of case reasoning: Roman Catholic casuistry of the High Middle Ages and the Renaissance. They present us with two arguments in one. First, moral philosophy is inadequate for resolving controversies and moral knowledge and justifications are particular and concrete in a way that theoretical knowledge is not. To resolve moral controversies one must use a method of resolution that is particular and content-full. Second, they argue that casuistry, modeled on the casuistry of the Roman Catholicism of the High Middle Ages and Renaissance, is such a method, and it can be deployed into the contemporary secular world.

Jonsen and Toulmin begin by arguing that moral knowledge is best understood as practical knowledge; that is, as knowledge that is concrete, temporal, and presumptive[4] rather than as a form of theoretical knowledge. Moral knowledge is particular to circumstances, times, and places. From their perspective Western moral philosophy has mistakenly understood moral knowledge as theoretical; that is, as knowledge that is idealized, atemporal, and necessary (e.g., geometry).[5] Foundational arguments are proofs that aim at knowing truly, while practical arguments are methods for resolving problems.[6] Theoretical knowledge and methodology do not depend on the circumstances or the situation of the knower. In setting out their understanding of the nature of moral knowledge, Jonsen and Toulmin follow an Aristotelian view that moral knowledge is a species of practical rhetorical knowledge.

While medicine blends both theory and practice, clinical medicine is an example of practical knowledge. It deals with particular individuals, problems, and details of cases and real situations.[7] Jonsen and Toulmin argue that judgments about moral issues involve attention to details and knowledge of particular cases in a process of reasoning analogous to medicine.[8] Assuming that their argument about the nature of moral reason is true, casuistry emerges as the best form of practical reasoning for the moral life. However, it is good to bear in mind that while they look at the issues of clinical medicine and find the need for casuistry, others, like Pellegrino and Thomasma, examine the same reality and see the need for virtue.

Jonsen and Toulmin support their argument by examining two very different public moral debates: the use of human subjects in research and abortion. They present the discussion of these issues as illustrative of two different approaches to moral judgment. Abortion is debated with "high theory and general principle."[9] This approach to moral problems assumes that "an ethical position always consists in a code of general rules and principles."[10] The foundational methods and principlism method, however, oversimplify the complexity of moral issues. In the casuist view, the method of theory and rules leads to a standoff between different theoretical positions, which, they point out, is the fate of the abortion debate. Each theoretical position captures some dimension of moral "truth" but each is incomplete. They argue that a consequence of such theoretical standoffs is ethical relativism. As Jonsen and Toulmin see it, if principles are interpreted absolutely, we end up with relativism since moral theories and principles lead to an intractable conflict in which one theoretical view or set of principles seems as good as another.[11]

The difficulty, as they present it, is a method that views moral knowledge as theoretical knowledge leads to a choice between absolutism and relativism. How, then, is bioethics to navigate the Scylla and Charybdis of absolutism and relativism? They think there is an alternative, exemplified in the work of the National Commission for the Protection of Human Subjects of Biomedical and Behavioral Research. On the Jonsen and Toulmin account the work of the commission began with concrete cases and problems, not with theories or general principles. The commission's work, which they see as a form of casuistry, was able to avoid both a theoretical impasse and complete relativism.

Jonsen and Toulmin see casuistry as "a reasonable and effective set of practical procedures for resolving the moral problems that arise in particular real-life situations."[12] Jonsen and Toulmin claim that casuistry redirects moral discourse from the theoretical level to that of concrete cases and to the principles and maxims people invoke as they face moral dilemmas.[13] At this least abstract level, we can secure agreement and avoid the endless debates about moral theory. By working at the case level the National Commission was able to reach agreements about human research and experimentation even though its members held diverse theoretical views. Jonsen and Toulmin repeatedly invoke the agreement of the members of the National Commission as an example of secular casuistry. These claims need to be carefully assessed.

For clarity's sake it is helpful to separate the different matters at issue. Jonsen and Toulmin raise very important questions about the nature of moral reasoning and justification. And the questions they identify are symptomatic of the postmodern dilemma in ethics. Moral pluralism raises questions about the content of one's moral views and about differing accounts of moral reason and justification. Both what counts as "evidence" (content) and rules of "inference" (method) are open to discussion. The second part of this book will turn to the matters of moral content and method. For the moment it is important to note that these questions are being raised at a primary level in bioethics.

However, the primary focus for this chapter is the method of casuistry that they put forth and whether or not such a model can be transferred to a secular context. While Jonsen and Toulmin are inspired by the work of the National Commission, they devote the bulk of their book to tracing casuistry from its ancient Greek and Roman roots to Pascal's damning critique in the seventeenth century. From this history the authors speak in generalities about a method of moral analysis, which they endorse for our use. In order to understand their method, its strengths and weaknesses, one must begin with their history.

JONSEN AND TOULMIN'S HISTORY

Jonsen and Toulmin argue that the casuistry of the High Middle Ages and Renaissance finds its beginnings in Aristotle, who distinguished practical knowledge (techne) and wisdom (phronesis) from theoretical knowledge (episteme). Moral judgment is one kind of practical wisdom while politics and rhetoric are others. For Jonsen and Toulmin, Aristotle faced the same difficulties in accounting for moral knowledge that we face today. Aristotle was trying to avoid the moral absolutism of Plato and the moral relativism of the Sophists.[14] While Aristotle did not, in their view, develop a casuistry, his thought forms the basis for the development of casuistry in later ancient thought. Aristotle took moral deliberation to be rational although not theoretical. Moral arguments are rhetorical in that they address a practical consideration, a moral controversy,[15] and are not deductive proofs; rather they seek to move people to action. According to Jonsen and Toulmin practical case reasoning was further developed in the West by the Romans, particularly the Stoics, Cicero, and the practitioners of Roman law. The Stoics sought to reconcile the demands of the natural law of reason with the demands of everyday life. In attempting to reconcile reason and everyday life, the members of the Old

Stoa used cases for instruction about the moral life; that is, to live, in the absence of passion, according to the Reason of the universe.[16] However, it may be worth noting that Jonsen and Toulmin point to a rule, not a case, in discussing Stoicism.

The context of the Stoic natural law provided the formative background for Cicero's development of casuistry. For Cicero the natural law of reason and the moral claims of social relationships (humanitas) must be worked out in relationship to the particular circumstances of each case.[17] For Jonsen and Toulmin the third book of Cicero's *De Officiis* is the cradle of casuistry. It is there that the nature of morality is debated in terms of theoretical arguments and specific cases.[18] In their view, the casuistry of the ancient world concentrated on various issues, introduced maxims and arguments relevant to each, and moved toward closure and judgment.[19]

Jonsen and Toulmin also see examples of case reasoning present in the early Christian tradition. This history begins with the letters of Paul and the writings of the Fathers of the church[20] and continues into the penitential books gradually developed for use by the clergy in the sacrament of confession. Such books reflected concerns about the particular circumstances of the person and the sin which confronted the confessor. In some of the penitential books one finds the comparison of the confessor to a physician. For example, the *Bigotan Penitential,* of the late seventh century, suggests that the power of a confessor should increase just as the power of a physician should increase "to the degree the illness of the patient increases."[21] Jonsen and Toulmin describe the penitential books as a "seedbed" for later casuistry.[22] Jonsen and Toulmin refrain, however, from seeing these books as presenting one form of casuistry. This is important for their understanding and presentation of the history of casuistry and their hopes for contemporary, secular casuistry since different penitentials represented different casuistries and practices.

The practice of casuistry in the High Middle Ages and Renaissance had a uniformity because it was situated under and controlled by ecclesiastical authority. If one saw, in the penitentials, a genuine practice of casuistry, then one would be open to a plurality of casuistries just as there were a variety of casuistries in the penitential practices.

Jonsen and Toulmin sketch out the development of a Christian West in which there was codification of canonical, legal, and confessional practices in the life of the church. This centralization began with Pope Gregory VII (1073). It peaked with the pontificate of Boniface VIII (1302), who, in *Unam Sanctam,* declared that all temporal and spiritual power resided in the Roman Pontiff by divine decree.[23] A prime instrument in the centralization process was the use of a Roman legal model to organize a taxonomy of moral cases. Jonsen and Toulmin point out that decision making at this time emphasized the circumstances of each case and a belief in a common morality founded upon the natural law.

Casuistry was further transformed by the development of a theoretical context which framed it and by the sociological context in which it was practiced. Morality became part of the study of theology, which was considered the most prestigious of all intellectual disciplines.[24] Moral theology was shaped by a Western Christian view of the natural law guiding moral reasoning in promoting the protection of basic human needs: self-preservation, preservation of the species, a search for understanding, and the achievement of social coherence.[25] Casuistry was the application of the moral law in addressing particular circumstances. Natural law doctrine gave the medievals and Latin Catholics a strong system of principles, maxims, and rules. These principles, set in an understanding of how the world is structured and known, provided the context for the moral deliberations of casuistry.[26]

Jonsen and Toulmin present the years 1556–1656 as the period of High Casuistry. Casuistry developed at this time in response to conflicting principles and obligations of natural law theory and the new problems of the Reformation (e.g., Did a Catholic owe allegiance to a Protestant ruler?). In their generalizations about the practices, they recognize six constitutive elements of medieval casuistry. First, there is the use of paradigm cases and reasoning by analogy. Second, there is the use of moral "maxims" to specify principles. Third, there is an attentiveness to the details and circumstances of a case. Fourth, there is an understanding of the moral opinions that may or may not be entertained.[27] Fifth, arguments are seen as cumulative; that is, one looked to the history of the community to learn about a type of case. Finally, there was a resolution of the cases.

It is important to note that in outlining these steps Jonsen and Toulmin speak of casuistry in its relationship to the health of the institutions in which it was practiced. This is a point they leave underdeveloped in that they fail to come to terms with casuistry as part of a way of life.

THE HOPE

Jonsen and Toulmin contend that casuistry can be revived and deployed in a secular age and still resemble its medieval counterpart in substance and method.[28] They write of seven features that characterize the casuistry of the Catholic Middle Ages which will also be found in a contemporary casuistry:

1. Similar type cases ("paradigms") serve as final objects of reference in moral arguments, creating initial "presumptions" that carry conclusive weights, absent "exceptional" circumstances.
2. In particular cases the first task is to decide which paradigms are directly relevant to the issues each raises.
3. Substantive difficulties arise, first, if the paradigms fit current cases only ambiguously, so the presumptions they create are open to serious challenge.
4. Such difficulties arise also if two or more paradigms apply in conflicting ways, which must be mediated.
5. The social and cultural history of moral practice reveals a progressive clarification of the "exceptions" admitted as rebutting the initial presumptions.
6. The same social and cultural history shows a progressive elucidation of the recognized type cases themselves.
7. Finally, cases may arise in which the factual basis of the paradigm is radically changed.[29]

The paradigm cases establish the boundaries and terrain of moral reasoning. Resolving a moral question depends upon where it is situated in the terrain.[30] But the presumptions created by the cases are open to change, rebuttal, and development.[31] In some cases more than one paradigm can be applied.[32] Moreover the moral geography set out by the paradigm cases will have collective as well as personal dimensions, and the collective aspects will help to clarify the terrain.[33]

In this context Jonsen and Toulmin offer what appears to be their strongest argument for why casuistry is the model of moral knowledge. They argue that the nature of moral knowledge is particular, hence, casuistry is unavoidable.[34] They present the history of a very particular model of casuistry and suggest that it, or one analogous to it, can be deployed in contemporary ethics as in the deliberations of the National Commission.

────── THE CONTEXT OF CASUISTRY ──────

For a casuistry to work it needs to introduce some mechanism by which cases can be identified, discussed, and resolved. Baruch Brody, for example, deploys a method of intuitionism to develop his model of casuistry. Jonsen and Toulmin use an appeal to paradigm cases. They hold that a contemporary casuistry will resemble its medieval, Catholic counterpart in *both substance and method* and that such a method allows us to sidestep the problems of moral pluralism.[35] However, they fail to understand that method and content are inseparable.

One fundamental issue they must resolve is how to identify the set of paradigm cases in a morally pluralistic society. There are key elements in the matrix of casuistry that would be difficult to transpose to contemporary society. Attending to the disciplinary matrix of confessional practice makes manifest that a casuistry of paradigms cannot be developed easily for a morally pluralistic society. A morally pluralistic society such as ours will have many different paradigms to direct our reasoning about issues such as abortion, suicide, reproduction, sustaining life, or the allocation of resources.

Jonsen and Toulmin speak of the crucial role of paradigms, cases, and taxonomies for casuistry. Their hope is that moral controversies can be resolved by the guidance of a taxonomy of paradigm cases. The crucial difficulty faced by a secular casuistry on the model they have chosen is how to select the correct taxonomy of paradigmatic cases. In a morally pluralistic society there will be different moral views and different sets of paradigm cases. Furthermore, a secular society lacks the common social structures of moral authority which are crucial to the casuistry of the Renaissance.

While turning to the history of casuistry my goals are philosophical, with implications for secular bioethics. A history of casuistry aids us to think philosophically about what a moral "case" is and to grasp the implications of moral pluralism for controversy resolution in secular ethics in general and bioethics in particular. To illustrate the essential embeddedness of medieval casuistry in the moral sense and social structures of its time, I will focus on three elements in Jonsen and Toulmin's account: paradigms, cases, and taxonomies. The first challenge is to determine what the term "paradigm" means for casuistry. The term, as we know, has a wide variety of meanings and Jonsen and Toulmin do not dispel its ambiguities.

CONFESSIONAL PRACTICE AS A DISCIPLINARY MATRIX

There are at least five ways in which a paradigm can function. These different functions give rise to the different senses of "paradigm": metaphysical paradigms, construct paradigms, sociological paradigms, exemplars of knowledge, and exemplars for action.[36]

Richard Grandy notes that part of the confusion over paradigms stems from the use of the term to identify both a "disciplinary matrix" and one specific part of the matrix.[37] A "disciplinary matrix" identifies the web of symbolic, metaphysical, epistemological, axiological, and practical components that shape the research of a scientific discipline. Catholic confessional practice in the Middle Ages and Renaissance was a disciplinary matrix and the practice of casuistry was embedded within that matrix. Casuistry was a method that depended on shared moral views and communal structures of moral authority to determine which paradigm cases should be applied. To remove the method from the matrix Jonsen and Toulmin will have to account for some shared morality and some structure of moral authority. If this analysis is correct, it makes no sense to talk of this model of casuistry outside of the disciplinary matrix of Catholic confessional practice or some analogous structure.

One can see in Renaissance confessional practice different elements of a disciplinary matrix. While metaphysical commitments may be implicit in a disciplinary matrix,[38] in Catholic moral theology and confessional practice they were explicit. Thomas Aquinas, for example, discusses action in the language of "substance," "accidents," and "essence."[39] The metaphysical assumptions of Catholic moral theology were closely supported by epistemological assumptions in terms of which moral controversies could be identified and understood by analogical reasoning. The disciplinary matrix of confessional practice not only had metaphysical and epistemological presuppositions but it had axiological dimensions as well.[40] It is within this matrix that casuistry described by Jonsen and Toulmin was practiced. Roman Catholic casuistry relied on both metaphysical and epistemological presuppositions which enabled casuistry to compare cases. These assumptions made the concept of analogous reasoning a central intuition of the medieval world. Absent these background assumptions it becomes unclear as to how casuistry is to discover and organize the cases. Jonsen and Toulmin, however, do not seem to draw out the deep connection between casuistry and the disciplinary matrix of Catholic confessional practice, a point to which I will return.

Cases and Moral Commitments

Exemplars are cardinal components of a disciplinary matrix. Shared examples guide those who use a matrix and they bind its elements together.[41] Jonsen and Toulmin are correct to argue that cases are crucial in the practice of casuistry that they outline. However, the Jonsen and Toulmin account does not develop the cultural and social roots of casuistry. Paradigms function within a sociological context and while they are defined by this context they also help to support it. Different moral commitments, or rival points of view, support the variety of casuistries one finds in the ancient world. For example, if one looks at the cases which were of concern to either Paul or Augustine, one can see the influences of social and moral values that raised issues for Christians that were not issues for the casuistry of Roman law. Paul's concern about dietary laws or Augustine's concerns about military service for the Empire were issues raised by the fundamental value commitments of the Christian religious community. These were not cases that were common for the casuistry of the Roman law or Roman Stoics.

The prospect of military service, involving the swearing of oaths and the use of violence, was a moral problem for the ancient Christian but not for pagans. This is because one finds in Christianity a strong presupposition against the taking of life and the swearing of oaths. These circumstances challenge Christians about taking part in the military. In the West, it was not until theologians like Augustine gave criteria for the just use of force, properly exercised only by the state, that Christians could engage in military service.[42]

If one examines the *cases that were not shared* by other communities of the ancient world, one can describe the ancient world as having a pluralism of casuistries. The Jonsen and Toulmin account does not explore the social and cultural dimensions of casuistry. If one attends to these dimensions one comes to understand the often radical pluralism of the ancient world and the more tempered pluralism in the casuistry of the Christian West. In light of the social and cultural influences, if one understands the secular world as consisting of communities with different moral values and no moral authority to determine proper paradigm cases then a method of secular casuistry will be limited from the start.

Conscience and Cases

Medieval and Renaissance casuistry developed in relationship to Roman Catholic confessional practice. Its goal was to aid the confessor in guiding a penitent with a troubled conscience. In its most general sense, conscience identifies the disposition of the mind of one making a moral judgment. The disciplinary matrix of Roman Catholic confessional practice helped the priest diagnose the character of the penetitent's conscience (e.g., well-formed, scrupulous, etc.), as well as characterize the cases that troubled the penitent.

Moral controversies arise because of conflicting views of the moral world. Cases of moral controversy do not simply announce themselves as moral cases. They become apparent as such against a background of moral assumptions shaped by rival moral outlooks. For example, one can imagine a person confiding a dilemma of conscience to another and the confidant responds "So what's the problem?" Contemporary difficulties in establishing public policy about abortion, physician-assisted suicide, the use of fetal tissue, or pornography exemplify how the values held by different groups shape moral controversy. Abortion[43] is described in a wide variety of ways that range from an exercise of a woman's rights to choose to the act of murder. In morally evaluating physician-assisted suicide the cases can be described as acts of compassion or acts of mercy. People may judge pornography to be morally offensive for very different reasons; for some it offends women's rights and is part of the social oppression of women while for others the moral issue centers on illicit passions. In each of these areas different descriptions derive from different moral worldviews.

There are at least three types of moral problems that constitute "cases." First, there are moral dilemmas when there is agreement on the description of the case but no clear resolutions of it. The dilemma comes about because the case supports a variety of conflicting resolutions. The following situation can be described as a moral dilemma. An eighty-seven-year-old woman with some dementia requires a feeding tube because she has a GI blockage that does not allow her to eat. The woman resists any tubes. She pulls them out and repeatedly says that she does not want any type of tube inserted in her. To be kept alive she must be restrained. The medical house staff, the nursing staff, and the family are divided about what to do. They agree about the description of the case and the various options but they are just not sure which treatment, or nontreatment, option ought to be pursued. This is a moral dilemma.

A second type of moral dilemma is one that involves arguments within a moral community. In the controversy there is disagreement about how to identify the moral problem. Different criteria for identifying a moral issue are applied. The Roman Catholic discussion of birth control is an example of such a controversy. The dispute represents an argument within the tradition about how the "natural law" is best understood. Here there has been an ongoing dispute about appropriate descriptions of these issues. Is this a case of responsible moral love or an act of selfishness? Often a resolution depends on the structures of juridical authority within the community to bring closure. There are often pluralism and argument within a moral culture. However, such pluralism can be understood and examined because the members of a community share enough in common. Of course, the methodological concern for the contemporary secular bioethics is how much is shared in common.

Finally, there are moral cases that are controversies which involve agents from different moral communities. Such dilemmas lack not only common moral criteria but the structures of moral authority as well. Such differences are one of the distinguishing characteristics of bioethics in contemporary secular society. The continuing controversy in civil society over abortion exemplifies this third type of moral controversy. It is to this type of dispute that Jonsen and Toulmin hope to bring the practice of casuistry. The medieval model of casuistry is controlled by a communal moral sense and a common moral authority. It addresses cases of the first and second type. Indeed much of the work of casuistry is to help penitents resolve moral dilemmas brought to a confessor.

Renaissance casuistry relied upon more than a common moral view. It presupposed structures of communal authority to interpret and implement its moral sense. One's membership in the community, indeed one's salvation, depended on the proper resolution of moral controversies. For the Renaissance Christian cases of conscience were resolved by an appeal to the authority of Scripture (e.g., the Decalogue), Tradition (e.g., the Didache, the virtues, commandments of the church), and the demands of everyday life. The function of the authority structure was to determine which set of criteria should be applied to the case. In the practice of casuistry there is an element of procedural closure whereby resolution is reached by the juridical authority of ecclesiastical structures. The practice of casuistry became more and more uniform as the authority which formed conscience became more and more centralized.

The three types of cases illustrate the different roles of authority in this model of casuistry. Intrapersonal cases were most often addressed by the authority of the confessor. The confessor, in the role of judge, gave advice and penance to bring about both punishment and reform of the sinner. For example, in advising a penitent who was guilty of theft, the confessor would instruct the penitent to make appropriate restitution for his sin. On this model the confessor is a person *in authority* by virtue of his ordained office. Like the policeman the confessor can resolve disputes because of the office he holds.[44] In other cases the confessor is also *an authority.* For example, Sister Mary Rita, the Superior of St. Kevin's Orphanage, was asked by one of the orphans about the identity of his parents. Sister brought the matter to her confessor. Under one description there is a natural bond which the child should know. Under another description revealing such knowledge to the parents and society is to break a bond of confidentiality.[45] As an authority the confessor might ask a variety of questions to understand the strength of the child's request or the nature of the promise made to the biological parents. In so doing he would attempt to describe the case in a new way to lead to a resolution of the problem.

One sees the same two roles for authority in interpersonal, or communal disputes, such as the Roman Catholic discussion of marriage annulments or usury. In these cases the authority is that of the magisterium, or theologians functioning as *an* authority, or the judicial structures of the church functioning *in* authority.

Authority, Paradigms, and Taxonomies

Jonsen and Toulmin speak of paradigm cases as related to one another in a taxonomy or classificatory system that, historically, is often organized around the Ten Commandments or the virtues. By use of this system casuistry is able to identify moral disputes and work toward their resolution. Cases of moral controversy are resolved, in part, by situating them in this structure which was embedded both in moral theology and the juridical authority of the church.[46]

The medieval and Renaissance practice of casuistry took new shape when, in the thirteenth century, the Fourth Lateran Council declared that all members of the church (all Christians) must practice the sacrament of penance annually.[47] In moving to a universal and uniform practice of confession Catholicism moved away from the penitential books with a variety of penitential practices[48] to confessional manuals with accompanying systematic commentaries.[49] A universally mandated practice, with common instruction guides, was supportive of and supported by an encompassing system of social control.[50] The decrees mandated steps by which the local churches were to make it possible for the faithful to comply with the law. Bishops were to educate and supply suitable priests. They were also to have a Master of Theology at each cathedral. The decree radically reshaped Catholic moral theology and gave birth to the casuistry of the High Middle Ages and Renaissance.[51]

Throughout the Middle Ages one finds three different centers for moral theology. The first dominant institution of ecclesiastical and social life in the early Middle Ages was the monastery. In this setting the principal literary products of monasticism were the penitential books influenced by monastic asceticism. The second centers of theology were the cathedral schools in Europe's new urban centers. Following Lateran IV, these schools developed the *Summae Confessorum,* which were learned, systematic, theological treatises for framing the discussion of cases. The *Summae* were pastoral handbooks that attempted to guide confessors. They were quite different from the penitential books insofar as they brought a more systematic ordering and discipline to the practice of casuistry.

Moral theology's relocation from the monastery to the cathedrals represents at least two important developments. First, the move represented a shift from a monastic spirituality of indifference toward the world to a spirituality that reflected engagement in the world and responsibility for society. Second, the move toward the cathedrals represented a move toward the centralization of authority in the episcopacy. The sacrament of confession was administered more often by secular clergy who implemented the more centralized and codified legal structures of the church and moral theology.

The third and final phase in the development of medieval moral philosophy and theology relevant to an account of casuistry began in the thirteenth century with the development of the university and academic theology. The university sought independence from both civil and episcopal authorities—engendering a tension which still exists today.[52] Episcopal control continued over the sacrament of confession and moral theology by the establishment of seminaries as schools, under episcopal control, to train confessors.[53] In particular, at the time of the Reformation, the Council of Trent established seminaries and their curricula using the current university models of education. Simultaneously the council reenforced the authority of the confessor over the penitent by describing priests as the proper ministers of the sacrament, as vicars of Christ, to whose judgment, in the sacrament, all Christians must submit.[54]

Jonsen and Toulmin do not develop the intimate relationship between casuistry and moral theory. In the late Middle Ages and Renaissance casuistry was both a way to develop and teach moral theology. The history of moral theology and confessional practice shows clearly that the practice of casuistry was embedded in a particular communal context defined by a complex set of moral assumptions. In their hope to develop a modern, secular casuistry they speak of medieval and Renaissance casuistry as if it was not shaped or informed by and reflective of a principled moral vision.

Casuistry was an attempt to guide the troubled consciences of Christians struggling with the moral demands of the world and the teachings of the church. It was also a way to control behavior. Some argue that Lateran IV mandated the practice of annual confession as a way to stem the spread of the Albigensian heresy.[55] The moral casuistry of the late Middle Ages and Renaissance was part of a communal practice embedded in and supported by metaphysical, epistemological, and axiological assumptions as well as by the authoritative structure of the episcopacy.

In the history of casuistry one can see the dilemma for contemporary secular casuistry. The history is that of a movement from the penitential books and practices to a more systematized casuistry. The pluralism of the penitential book was brought under the control of a greater uniformity by the ecclesiastical authorities. In a contemporary secular setting, with greater diversity and no common moral authority, there is an even greater likelihood that there will be many casuistries.

Confessors and Casuistry

Confessors played a central role in the practice of casuistry, to the extent that they selected and applied paradigms to the cases before them. Lateran IV prescribes:

> The priest is to be discerning and careful, so that like a skillful doctor he can apply wine and oil (Lk 10/34) to the wounds of the injured person, diligently asking for the circumstances of the sinner and of the sin, through which he can *prudently* understand what advice he ought to give, and what sort of remedy to apply trying various things to heal the sick person.[56]

The role of the "prudent" men who were to be confessors is crucial for the practice of casuistry.

"Prudence" has a much different meaning for the Renaissance Christian than for the person of the modern world. Understanding the different senses of "prudence" is important for understanding this method of casuistry. In a contemporary understanding "prudence" is frequently seen as rational self-interest. Kant understands prudence as "cunning" and views it as evil. The medieval and Renaissance understanding of prudence was grounded in the Aristotelian virtue of *phronesis,* which was an intellectual virtue required by the other virtues.[57] It was not moral wisdom but the ability to be practically wise.[58] However, to be morally virtuous one needed to be practically wise[59] and the man of practical wisdom is one who is virtuous.[60] Phronesis is the exercise of right choice in particular cases, *in light of more universal knowledge.*[61] Cicero translated the Aristotelian term phronesis as "prudentia" and spoke of it in moral terms.[62] For Aristotle phronesis was tied to a *paideia;* that is, a culture, a civilization. For Cicero phronesis was grounded in a view of *humanitatis.* For Saint Thomas prudentia was one of the four cardinal virtues. It was "knowledge of what should be done and what avoided" and guided reasoning about what ought to be done.[63] As a "special virtue" for Saint Thomas, prudence had elements of both the moral and intellectual virtues. It was understood as a faculty for applying universal principles to particular problems. Prudence does not appoint ends but regulates the means to the ends.[64]

It is interesting to note that for the Jesuits, some of the most noted practitioners of casuistry, prudence was one of the most admirable virtues. In the language of Saint Ignatius, it is the virtue of "discerning love." Ignatius uses the language of "prudens caritas" and "discreta carita" interchangeably in the Constitutions of the Society of Jesus.[65] His usage of "prudence" and "discretion" connotes the language regarding the discernment of spirits developed in his *Spiritual Exercises.*[66] Prudent agents are guided in making choices by discrete charity which impels them to choose the better course after all the circumstances have been considered. The root of this discrete charity is the interior law of charity described by Paul in his "Letter to the Romans."[67] The presence of the Holy Spirit gives the agent the grace to carry out what the law requires.[68] In the practice of casuistry the confessor needs more than just knowledge of paradigm cases, he must also be prudent in finding the best paradigm for the case and individual before him.

As one recovers the role of prudence in the practice of casuistry, one can grasp the centrality of a particular moral vision and sensibility that was and is crucial to the practice of Catholic casuistry. The contextual embeddedness of medieval and Renaissance casuistry occasions asking whether, in a secular age fragmented by many different moral senses and eschewing the guidance of the Holy Spirit, casuistry can be practiced on a medieval model.

Auxiliary Principles

One can glimpse in yet another way the cultural embeddedness of Renaissance casuistry by examining two of its crucial auxiliary principles: the principle of double effect and the principle of probabilism. The former reflects casuistry's struggle to address difficult cases from within a moral framework; the latter addresses questions of proper authority.

The principle of double effect, central to many discussions in bioethics and rooted in Thomas's treatment of self-defense, is a staple of casuistry.[69] The principle is primarily used in three types of cases: scandal, killing, and sexual cases. The development of the principle reflects a particular set of moral assumptions: (1) certain acts are *mala in se* and (2) the intention of the agent defines the morality of the action.[70] If one does not hold these assumptions, the principle of double

effect becomes unnecessary to a consequentialist or a laxist manipu-
lation to a deontologist.

Jonsen and Toulmin note that casuistry has the ability to respond
to changing situations, social circumstances, and practices. They describe this adaptive process as one of clarification and elucidation.[71] In
the medieval practice of casuistry, probabilism emerged as a central
way to resolve emerging moral dilemmas. Probabilism holds that, in
a case of practical doubt, a probable opinion may be followed even when
the contrary opinion is more probable. It employs the principle that a
doubtful law does not oblige, a principle articulated by Bartolomeo
Medina in 1577.[72]

The discussions of probabilism were initiated in areas where opinions concerning cases were in conflict. But most importantly, probabilism developed to relieve the consciences of men and women from
the burden of too grave an obligation. In its development three central factors emerged: (1) the role of conscience, (2) the role of authority, and (3) their relationship one to the other.

In making moral decisions, the medieval and Renaissance mind
had to reconcile a number of sources of moral authority: (1) codes of
law and sacred books (e.g., the Scriptures), (2) opinions of individual
experts concerning morality (theologians), and (3) those in authority
(e.g., confessors, bishops). One can view the discussion of probabilism, so important to medieval casuistry, as a discussion of proper authority.

There were several different approaches to situations when it seemed
as though an obligation might exist, but the matter was not perfectly
clear. "Probabiliorism" held that when there were arguments in favor
of an obligation and others in favor of freedom, one ought to follow
the obligation. "Equiprobabilism" attempted to effect a compromise
in that when obligation and freedom seemed equally balanced one
could morally follow the option of freedom. But where the opinion
of freedom seemed to be least likely one ought to follow the more
rigorous course.[73]

Gabriel Vasquez characterized opinions as *intrinsically* probable if
founded on excellent argument, or *extrinsically* probable if founded on
an opinion of a wise man.[74] The authority, learning, and prudence of
others were taken as proof that the opinion in question was probably
a correct opinion.

Another element of probabilism came from Suarez, who developed what became known as the "practical principles" to help achieve practical certitude. His two principles of possession and promulgation had to do with conscience and authority: "in doubtful matters, presumption is in favor of the possessor, and a law does not bind unless adequately promulgated."[75] In cases of genuine doubt about an obligation (where an authoritative magisterial opinion had not been promulgated), a person could arrive at a judgment of conscience. Suarez's principles aided in the exercise of conscience.

The Suarezian principles, along with Vasquez's principles of the extrinsic and intrinsic authority, illustrate the interplay of the structures of communal authority and a theological framework that defined the fabric of medieval casuistry. Returning to distinctions drawn earlier, probabilism sought to address areas where the extent of one's moral obligations was not clear. Since different "authorities" held different opinions, the doctrine sought to sort out, for members of the community, a justification for following the opinion of one who was *an* authority.

The importance of authority in the life of the community may be difficult to grasp for those raised in a post-Enlightenment, democratic culture. Much of the modern philosophical project has been a search to justify "authority." One finds in morality, politics, science, and epistemology the effort to secure an authoritative grounding for each discipline. In this quest for authority the individual person takes on central importance as the source of authority and authenticity. The modern age does not appeal to God, revelation, or tradition for public authority. It has sought authority in other areas (e.g., reason, moral sense, autonomous moral agents). In this search to ground public authority the individual person has taken the role as the center of authority.[76]

After examining the elements of casuistry which Jonsen and Toulmin highlight, one cannot help but conclude that their hope to appropriate the methodology of medieval and Renaissance casuistry for a secular, morally pluralistic world is freighted with difficulties. The practice of casuistry was embedded in complex cultures with understandings of moral authority that directed the practice. The broad parameters of the casuistry were articulated in moral theory of natural law and an account of the virtues. While there was no doubt pluralism within the interpretations—one has only to think of the disputes in moral theology between the Dominicans and the Jesuits—there were boundaries that confined the disputes. The practice of casuistry benefited not only from an account of the virtues but also from role models for moral behavior. Each element—culture, authority, moral theory, saints, and heroes—influenced and supported the practice of casuistry. It is not clear how casuistry would have functioned without these supporting elements.

AN EXAMPLE OF RENAISSANCE CASUISTRY

Jonsen and Toulmin provide three examples of the practice of casuistry (usury, perjury, and pride). Their examples serve to illustrate the difficulties of applying this model of casuistry to a secular, morally pluralistic culture.

Jonsen and Toulmin trace the origins of the Christian prohibition against usury to scriptural, councilor, and canonical prohibitions.[77] It was defined as a charge of *any* interest at all for a loan. Usury was considered a sin because of a scriptural passage and a theological interpretation of that passage. The Book of Deuteronomy (Dt 23:19–20) forbids the lending of anything for any interest to one's brother. It does allow loans, at interest, to foreigners. The Western Christian tradition, following Saint Jerome, understood all people to be "brothers" since Christ's salvific act was universal.[78] In the twelfth century usury was classified by theologians, such as Peter Lombard, under the heading of theft.

The casuistical paradigm began from that point and defined usury as "mutuum est quasi de meo tuum" (my property is made yours). This formulation recalled the practice of Roman law in which property was transferred when lent to a borrower. Again, the cultural heritage of the West was important in the shaping of the moral sensibilities about usury. The roots of the prohibition against usury were found in Roman law and in Christianity. The agent of the loan had no right to charge for use of the property, since during the time of use, the loaned property did not belong to the agent because what was lent had been borrowed by a "brother." The moral presumption was that one ought not make a loan, at interest, to one's brother.

Canonists and theologians found the prohibition against usury to be a part of the natural law. Thomas Aquinas added yet another qualification in the prohibition against usury in that he held that money was not saleable.[79] During this period, however, commercial activity was increasingly recognized as morally licit. People could form partnerships in which money was committed to a joint enterprise (e.g., societas). In such cases, it was held that the use of the money was not given over to another as in the case of a loan.

It was understood in the medieval and Renaissance world that all of the natural law admitted exceptions (except for the primary precepts). So too, the prohibition of usury had exceptions. The most important exception was the obligation of a debtor to make the creditor whole when the creditor suffered because of the loan. In such cases the interest was a moral and legal payment described in terms of compensation for damages. Social circumstances, from the fifteenth century onward, led to new understandings of time and to "new perceptions of the meaning of money and credit."[80] In these new circumstances the casuists brought together the notions of 'risk' and 'damage' to frame a new paradigm for describing usury.

The threads of casuistical thinking on usury were brought together by a Dominican, Johannes Eck, under the model of the "triple contract." First, there was a contract of partnership. Second, there was a contract of insurance. Third, there was a contract for a return rate of interest described as insurance in that the investor could have put the capital to more profitable and less risky uses.

The triple contract raised questions which, ironically, were posed by the German Jesuits in Ausburg—home of the Fuggar banking house. The commission which investigated the matter reported to the Fourth General Congregation of the Society of Jesus. The congregation gave what became the authoritative statement on the practice. It held the charging of interest to be morally licit since a probable opinion existed supporting it on grounds that in areas of undefined teaching a probable opinion establishes liceity. Central to the resolution to allow usury was the notion that in the contract the author of the loan suffered the loss of profit from the capital. By choosing to invest, the lender lost the direct benefit of the money during the period of the loan.

In examining the case of usury we discover the importance of the moral vision of the community in identifying paradigm cases. In the inception of the case and its resolution, one finds the moral commitments of the community and the authority of the community at work. The debate illustrates not only the role of communal authority, but it is also an example of a communal tradition responding to changing circumstances.[81] Both levels of the authority structure were involved. The theological experts debated the issue and those in juridical authority brought procedural closure to the debate.[82]

One sees in this historical example how casuistry was practiced in a tradition with a well-defined moral sense and the structures of authority to interpret that moral sense. The exercises of authority were not merely exercises of power but ways in which the moral sense of the community was reinterpreted to address the new challenges and changes in circumstances. One can describe this exercise as an effort to maintain a cohesive moral vision in the community.

As one contemplates moving the structure of medieval and Renaissance casuistry to the secular world, as Jonsen and Toulmin suggest, we will not only have to address the pluralism of moral visions but also the question of who constitutes moral authority and moral "experts." A morally pluralistic world will have many who are morally wise, though they hold dramatically different views. The shift away from structured moral authority and experts is illustrated in the Protestant Reformation. In general, casuistry falls out of favor because the Reformation view of conscience and authority is so different from the Roman, medieval view. The Reformed churches asserted conscience to be outside the regulation of any institution or community. The assertion of the independence of conscience led to a new understanding of appeals to authority and led to casuistry falling into disuse in the Protestant world. Even the Anglican churches reshaped casuistry by stressing the importance of the autonomy of individuals. In the Anglican churches casuistry continued to function primarily as a part of moral education.[83] With the shift toward the authority of the individual conscience casuistry took on a much less significant role in the life of the community. What happened in the Reformed churches illustrates, in part, the difficulties for a secular casuistry. With the locus of moral authority situated in the individual person casuistry is individualized rather than communal.

SUMMARY

For Jonsen and Toulmin one of the advantages of casuistry is that it changes our understanding of moral knowledge and allows us to speak of moral knowledge in contrast to deductive, theoretical knowledge.

They argue that theoretical knowledge can be spoken of in terms of "necessity" and "validity" as its elements are applied within a *system* of concepts.[84] They take practical moral wisdom to be analogous to the practice of clinical medicine in that it applies to a world of concrete objects and actual affairs.[85] In this, Jonsen and Toulmin have made an important challenge to our understanding of moral knowledge and justification. The goal of practical knowledge—be it moral wisdom or clinical medicine—is primarily to do something well. It is not the primary goal of practical knowledge to know truly. Of medical diagnosis Henrik Wulff has written: "It is meaningless to discuss if the resulting definition is true or false as all definitions are arbitrary, but it is possible to ask whether or not it serves a practical purpose."[86] On the casuistical account, moral reasoning resembles the work of the physician more than the work of the scientist. Indeed, the comparison of confessors to physicians, as practitioners of therapy, is a very early image in Christian literature.

The practice of casuistry was part of a way of life. They have attempted to excise a practice from its context and transfer it to a very different context. However, the reference points of a communal context no longer have the substance and glue of the practice. In this method of casuistry the identification of paradigmatic cases took place in order to solve the dilemmas of confessors and penitents. The goal was to enable the confessor to direct and relieve the conscience of the penitent. The casuistry Jonsen and Toulmin chronicle is set within a communal context that possesses moral and juridical authority as well as a broad moral vision. The casuistry deployed by Jonsen and Toulmin is embedded in axiological, metaphysical, epistemological, and sociological elements of the matrix of Roman Catholic confessional practice without changing the character of the practice and removing the original grounds and circumstances for it. The method of casuistry used by Jonsen and Toulmin had communal authority and a content-full moral view as necessary conditions for its practice. The method of casuistry, like other methodologies in ethics, cannot be separated from a particular content.

The importance of these conditions can be seen when one considers how difficult it is to develop analogies in a secular casuistry that lacks the communal moral sense and structures of moral authority. Think, for example, of cases about treatment at the end of life. For some the analogous relation of these cases is the intention of the agents, for others the analogous relation is the causal relationship between the physician and the death of the patient, while still others would focus on the exercise of patients' rights. While all might agree on the groupings of these cases, they are very different moral views.

The difficulties in transposing Catholic casuistry to the contemporary secular world are seen in the seven features that Jonsen and Toulmin regard as part of a secular practice of casuistry.[87] As one examines the seven features outlined by Jonsen and Toulmin it becomes clear that their hope is misplaced. Absent a particular content-full moral framework it is impossible to establish the paradigm cases which are to be the "final objects" in moral arguments. Absent the shared moral point of view not only will the paradigm cases be missing, but there will be no common criteria by which to identify and describe the moral controversies for which we seek paradigms. One does not have to endorse Engelhardt's view[88] that our world is populated by moral strangers in order to recognize that a method of casuistry, so embedded in a particular understanding of morality, cannot be imported to give like service in a morally pluralistic world. Even if we could somehow establish shared paradigms so as to overcome the problems of incommensurability, from a secular standpoint we would still lack the common social structures of moral authority needed for the ongoing development of casuistry modeled on that of the High Middle Ages and Renaissance.

The history of Catholic casuistry raises philosophical questions about the nature of moral "cases" and what it is to resolve such cases. Jonsen and Toulmin, unfortunately, seem to have missed these issues.

Earlier in the chapter I generalized Jonsen and Toulmin's argument as follows:

1. Moral reasoning, as practical reasoning, is best understood by thinking about practical, concrete, moral dilemmas.
2. They present a history of one model of case reasoning.
3. They suggest that such a model can be transferred to contemporary secular culture, and they support their claim by highlighting the work of the National Commission.

However, when one brings in other elements of the method one sees it to be deeply embedded in the values and structures of a particular moral community. On this historical account there are now good reasons to hold that this model cannot be transferred to a secular, morally pluralistic world. The history illustrates the importance of common moral views and communal authority in directing casuistical reasoning.

What are we to make against Jonsen and Toulmin's experience of the National Commission? It is possible to describe the workings of the National Commission in a way which is not at all casuistical. While the commission had a diverse membership, it is not hard to imagine that the members, nonetheless, shared a common background of moral values. A change in the membership (to include people like Robert Nozick or Paul Ramsey, for example) might have made for much different outcomes in the "agreements" reached by the commission. As chapter 5 will argue, case agreement can be reached if common moral commitments exist. The success or failure of secular casuistry will depend on the extent of agreement among moral intuitions. The ability to move from one case to another will also depend on the depth of agreement in the shared moral sense. Some form of casuistry may be possible, but it will be in a more limited form than the method for which Jonsen and Toulmin hope.

——— Secular Casuistry: ———
The Paradox of Trolleys and Transplants

Jonsen and Toulmin selected a very particular, well-defined method of casuistry in their attempt to offer a response to the search for a method in secular bioethics and moral philosophy. Casuistry is a method, I have argued, that will be limited in a morally pluralistic society since there are different moral views of paradigm cases and moral authority. One might respond to the criticisms advanced against the Jonsen and Toulmin account by saying that there are other models of casuistry which have been developed and which could respond to the needs of applied ethics in a secular world. However, this view does not take into account that any model of casuistry must establish a set of paradigm cases which in turn guides the practice of secular casuistry. To establish a method is to establish a rule, or set of rules, that are grounded in a way of life.

The problem confronting a contemporary practice of secular casuistry is how to select the paradigm cases and resolutions. There are two possible ways by which casuistry could identify paradigm cases. One way would be by some shared set of moral commitments, values, or intuitions. Another way for casuistry to work would be to rely on structures of authority which would identify, articulate, and resolve the paradigm cases and link other cases to the paradigms. Legal casuistry is an example of a model of casuistry which relies on a structure of authority. The practices of casuistry in the Roman Catholic community, the Jewish community, and the legal community as well are marked by casuistry that is shaped and defined either by the moral sensibilities and concerns of the community or by the structure of the community (e.g., law). The model Jonsen and Toulmin emulate is one that has both a particular moral sense and a well-developed authority structure to resolve controversies. Casuistry does not necessarily require the structure of the Roman Catholic model, but it does require either a shared moral sense or a process for closure and resolution. In our contemporary secular context there is no communal structure with moral authority to arbitrate and resolve moral cases unless they touch on matters of law. Other casuists, like Baruch Brody, recognize the lack of shared paradigm cases and rely on common moral intuitions to direct a model of casuistry. The success of the Brody method will depend, in large measure, on the extent of shared moral intuitions.

Many will argue that the discussion of the "Trolley Paradox" represents an example of successful secular casuistry. The discussion of the cases is important for my argument since some may think the cases provide a rejoinder to my conclusions that casuistry is not possible in a secular society.

Briefly, the cases of the paradox are:

Case One: A trolley comes down the track and as it rounds the bend there are five people in its path ahead who cannot flee. If it continues on its path, the five will surely be killed. It is possible that the path of the trolley can be switched to another track where there is but one person who, of course, cannot move in time to escape the oncoming trolley and will be killed if the switch is made. May you (as the driver or the switcher) change the direction of the trolley to the second track so that the one will be killed and the five spared?

Many who have puzzled over this case and its variations have held that one may change the direction of the trolley. The paradox comes when this case is examined along with the second case.

Case Two: You are a transplant surgeon with five patients in need of whole organs in order to live. You are such a skillful surgeon that the success of the surgery is assured. Each is about to die when a healthy, young man, who would be a perfect match for each of the five patients, walks into your office. May you kill the one so that you may take his organs and save the five? Most have said that one ought not kill the healthy young man in order to harvest the organs to save the five.

The paradox for those who have thought about these cases has been to explain why it is permissible to turn the trolley in case one and impermissible to kill the healthy patient in the second case. The literature on these cases is rich, complex, and puzzling. Rather than trace the details of different authors' reflections, it is more helpful, for my project, to examine what this literature contributes to a model of casuistry.

The writing on this paradox has focused on the cases rather than on an understanding of casuistry. This literature is instructive in a way the authors may not have intended; that is, as a way to see some of the elements of a secular casuistry. Throughout the many discussions of this paradox there are appeals to "intuitions" and a search for moral "rules" and "principles."

SHARED INTUITIONS

In the philosophical literature much discussion of the trolley-transplant paradox is cast in terms of "intuitions" and "feeling."[89] Such language indicates a necessary condition for moral casuistry to work. Those interested in resolving moral cases must have some basis of shared or overlapping intuitions with which to identify and resolve the case. Indeed, Francis Kamm speaks of the trolley discussion as going on among "[S]ome moral philosophers with *nonconsequentialist* leanings. . . ."[90] The general philosophical discussion has involved how to explain the intuitions and resolutions of the paradox.

The paradox points out both the possibilities and the limitations of a secular casuistry. The extent to which a secular casuistry will work depends on a coincidence of moral commitments. The ability of secular casuistry to resolve moral controversies depends on the extent to which moral commitments are shared. For example, one can return to the paradox and imagine a consequentialist, with different sets of intuitions, not finding much of a paradox with the two cases. A consequentialist who is a total or average utilitarian might say that the track with the single person on it should be taken and that the healthy man should be killed for organs. A rule consequentialist might hold that the trolley should go down the track with one man but that the healthy man should not be killed for improved overall consequences for the society. The rule utilitarian / consequentialist may find no paradox to speak of and argue that general utility is served by protecting the one man. Each of these utilitarians brings a different set of moral sensibilities to the trolley-transplant paradox.

One can describe the paradox by telling the story in yet another way; that is, from the point of view of the agent (the driver or the transplant surgeon) and his intentions. If one has an absolute constraint (e.g., against doing what is a *malum in se*), such as not killing innocent human beings against their will, one can arrive at the same judgment about what ought to be done but for very different reasons. The justification will be different from someone who does not hold an absolute constraint. How the story is told and resolved will be influenced by one's moral commitments. This illustrates the difficulties confronting a practice of secular casuistry in that, absent a common justificatory framework, it will be difficult to identify or resolve new controversies.

As Alasdair MacIntyre has pointed out, moral arguments often have an intractable quality to them. It is not that the arguments are invalid, for they often proceed by valid inference from the premises. The intractable quality is in the acceptance or rejection of the initial premises.[91] Or, one might think that the initial premises are like rules for resolving moral issues. But rules and premises come from within a way of life. One can see the possibilities and limits of secular casuistry if one recalls, in the literature on abortion, the famous case in which Judith Jarvis Thompson suggested that one could talk of the relationship of the fetus and mother as analogous to a violinist attached to and dependent on a person in order to live.[92] Thompson's analysis of abortion, by use of this case analogy, did not achieve the general agreement we find in the trolley-transplant paradox.[93] Just as initial premises set the framework for arguments, so too initial intuitions set out the groundwork for casuistical reasoning.

The violinist example is but one example of a case where people respond with differing intuitions. Regularly the *Hastings Center Report* publishes cases with commentaries. It is striking how different the commentaries often are. In these different commentaries one sees how moral perspectives shape the way the facts are arranged. Some facts are highlighted, while others are down-played and still others disappear.[94] Our moral perspectives are important to the way we understand and act in the world. For example, is the removal of artificial feeding and hydration from persistently vegetative patients a case of "starvation" or the cessation of useless medical treatment? It seems clear from the general literature of ethics, as well as the literature of bioethics, that one's moral perspective will shape the way cases are seen, described, and resolved. Those intuitions must be shared if casuistry is to work.

If, as I have argued, a method of casuistry depends on common moral commitments then, given disagreements over issues such as abortion, the treatment of persistently vegetative patients, the use of fetal tissue, or the distribution of health care resources, we are likely to have as much disagreement as agreement about what are cases of moral concern and the paradigms for resolving them. Secular casuistry will be successful to the extent that there is shared moral commitments that allow the identification of cases and their resolution.

Again we return to an earlier theme of what it is to have "agreement." The trolley-transplant discussions illustrate how people with diverse moral intuitions can reach agreement on a particular case because they really begin with agreement. The extent of agreement, however, will be crucial for the ability of secular casuistry not only to resolve the case but to extend resolution to other cases. Those times when there is "agreement" one must examine the extent of the agreement to see if it is more than superficial (see chapter 6). I think that the category of moral acquaintances can help understand agreement and disagreement about cases.

There is, however, a more profound question about shared commitments. Even if we found we were in a situation where there were common moral intuitions about cases, we still have to answer why we should take these particular intuitions to be normative. Are our intuitions expressions of ontological realities about the moral qualities of the world? The past four centuries of philosophy remind us of how difficult it is to understand such claims.[95] While moral facts may exist, the difficulty, for humans at least, is knowing what they are. If our intuitions are not rooted in real moral "facts" but rest on social practices, why should we give them normative weight? Our intuitions may be distorted by ignorance, prejudice, power, or a host of other influences.[96] We must be cautious in deciding the extent to which our intuitions have such normative force.

SUMMARY: TROLLEYS, TRANSPLANTS, AND DISAGREEMENT

It is clear that there are cases about which there is some common agreement (e.g., the trolley-transplant paradox). However, there exists a wide array of cases that elude paradigms or even common description (e.g., abortion, fetal tissue). The cases of agreement give rise to the hope that, absent theoretical agreement in ethics, moral agreement can be reached through some form of case reasoning. The cases of disagreement, however, give one pause. Those cases are unresolved because we lack common descriptions of the cases, and the cases will remain unsettled at least until we can reach common descriptions.[97] Moral friends should be able to share common descriptions of the cases. Moral strangers probably will not. And moral acquaintances will be able to share some descriptions but not others.

The practice of secular casuistry will be difficult, if not impossible, in a morally pluralistic world. Secular casuistry will only work when there is a coincidence of like-minded moral intuitions. The resolution of cases can also be achieved when the lines of decision making authority have been delineated and the decision maker(s) can be identified (e.g., free and informed consent). While a religious casuistry relied on the authority of apostolic succession, a secular casuistry may rely more and more on the authority of procedures.

—— Conclusions for Bioethics ——

Jonsen and Toulmin articulate a hope that casuistry provides a method of moral reasoning for bioethics in a secular context. The model of casuistry they selected, however, is limited for a secular context given its place in a defined moral sense and social structures. However, even a less structured model of casuistry will have limitations for bioethics in a pluralistic context. Casuistry can flourish within a moral community. Its ability to serve well outside a particular community will depend on an overlapping of moral commitments.

The principal challenge for any practice of casuistry in bioethics is to establish a common moral framework in which cases can be identified and resolved. What are the paradigm cases that will resolve controversies about abortion, physician-assisted suicide, genetic medicine, fetal tissue, or embryo research? That is, without shared moral values, sensibilities, or intuitions, we will not be able to resolve moral controversies or develop principles and rules from the cases. Moreover, we will not be able to even identify which cases are candidates for moral controversy without a shared moral sense. Developing a common description of a case will depend on common moral commitments.[98]

This, one might say, is a variation on the fact-theory and the fact-value distinction. Just as "facts" are not independent of linguistic expression, neither are "cases," nor their descriptions, independent of moral worldviews. The descriptions of the facts about the world are influenced by our values and vision of the world. What we see as important, unimportant, or troublesome will be influenced by our values and background assumptions. So too our descriptions of moral cases in need of resolution will be shaped by our moral values and intuitions. The possibility that any form of casuistry can resolve moral dilemmas will depend on the extent to which individuals share moral descriptions.

One way to describe the "success" of the National Commission is simply to say that the members of the commission shared enough similar moral commitments to guide their deliberations. The members may not have been moral friends but, at the very least, they were moral acquaintances. Again one has only to reflect, for example, on the controversies surrounding abortion, fetal tissue, and HIV screening to realize both the importance of our moral intuitions and the difficulties for resolving dilemmas when basic intuitions are not shared. One limit for the potential of an intuition-based method of casuistry is that it may well be limited to the cases at hand depending on the strength of the intuitions. Bioethics, however, addresses more than cases. It also works on areas of policy, law, and education. These other venues of bioethics may require other methodological choices.

Like the ancient world and the confessional practices of the early Middle Ages, there is a plurality of casuistries in the contemporary world. One can argue that medieval casuistry was pluralistic in at least two respects. First, it was practiced in other communities with different sensibilities and mechanisms than Latin, Roman Catholic casuistry. Second, in the early Middle Ages there were a variety of casuistries practiced within Christianity in the penitential books. Such a history of casuistry is more helpful for understanding case reasoning in ethics today where one will find, most likely, a variety of casuistries with agreements and disagreements about cases.

The methodological turn toward casuistry helps us to achieve an even better understanding of agreement and disagreement in bioethics. One of the paradoxes of the field has been that it has achieved a fair amount of agreement on important issues while remaining in profound disagreement on other issues. To better understand the moral acquaintanceship and the implications for method in bioethics, the topics of agreement and disagreement need to be further explored and analyzed.

PART TWO

Moral Acquaintances and Methodology

5. Communitarian Bioethics

The limitations of different methods in bioethics have led some in the field to search for a communitarian basis. The appeal to communitarian thought has been, in part, a reaction against a perceived over-emphasis on autonomy in the traditional liberal paradigms of bioethics.[1] In general, this method argues that the starting point for health care and bioethics ought to be the community rather than the individual. The argument generally runs that it is inappropriate, indeed incomplete, to consider individual human beings apart from the community. The communitarian approach supports the importance of autonomy but emphasizes that no autonomous person lives in isolation. The obviousness of the communitarian argument gives it an intuitive appeal. It is true that no human being exists alone. How women and men understand themselves is tied to the social and cultural context in which they live. Another part of the communitarian appeal is its recognition that morality is part of life. How we should act and what we should do are questions that are answered in the broader context of the meaning of life. The communitarian turn connects bioethics to these broader questions and decisions.

I view this turn as a positive step for bioethics. In it, bioethics is a systematic reflection on the moral life and on the moral life as contextualized in particular practices and ways of life. If, that is, the turn to community also recognizes that method and content are inseparable. To understand a method, one must understand both its context and content. Each method reflects assumptions and the commitments of different moral worldviews. It is hoped that by turning to the particular—the communal—we can better understand (1) the contributions and limits of the different methods in bioethics; (2) that which members of the secular society share in common; and (3) the roles a bioethicist can actually fill in a community and secular society.

It is important, however, to be clear at the outset that I will not argue here for a single communitarian bioethics. Moral communities come in a variety of shapes and sizes, and with different self-understandings and moral commitments. Since the communitarian method is embedded in the moral commitments of particular communities, it would be a mistake to argue for a single communitarian point of view or method for bioethics. One can only say that there are communitarian bioethics in the plural. One insight of postmodern thought is that we no longer have—if we ever did—a general moral narrative. In a postmodern perspective, the focus falls more on the contingent, the particular, and the local than on the universal and the necessary. In a secular society that is multicultural and morally pluralistic no single communitarian bioethics method is possible.

While noting the importance of the communitarian dimensions in an investigation of methodology it must also be recognized that men and women in modern secular societies are often members of a number of different communities. Various communities of family, faith, work, associations, and friendships shape the moral lives of most people, and these communities are not always consistent with one another. Having memberships in different communities can lead to the rootless cosmopolitanism described by MacIntyre.[2] Such diverse memberships can be a source of moral conflict. At times, different communities will make demands on a person that are inconsistent and incoherent. Communitarian bioethics must not overlook these differences or assume that all communities share the same moral commitments and methods. But the opposite assumption—that a person can only belong to one community—is also dangerous. Such a view would make certain assumptions about communal life that may be true for some communities but not for all. By understanding different models of community, one can better understand how collaborative and civic discourse emerges among different communities. The positive contributions a communitarian perspective brings to methodological reflections (e.g., an awareness of the social context of ethics and how it is part of a way of life) must be weighed in the balance with the limits of communitarian thought in a secular society.

The first section of this chapter argues that the turn toward community and particularity is inevitable for understanding method in bioethics, and one can identify a range or spectrum of moral communities. Most moral communities can be analyzed by three factors: the content of a community's moral vision; a community's understanding of moral authority and the offices of authority; and the community's self-understanding about how it should relate to other moral communities. In one sense, these are artificial distinctions and the different elements often blend together to form a unity of life in the community. However, for the sake of analysis and discussion these elements can be separated and, in fact, used to chart a wide spectrum of moral communities. At one end of the spectrum of moral communities is the "strong" community that possesses clear, and often detailed, views of the moral life. Such a community may have clear moral "experts" to define and properly interpret the issues. It may also have a strong sense of itself as distinct and separate from the world. At the other end of the spectrum are communities that have a common moral vision, but also a great deal of openness about that vision and its development. Such a community may have weak or democratic views of authority and an ecumenical understanding of its relationship to other communities.

—— PARTICULARITY, HUMANITY, AND COMMUNITY ——

There are good reasons for bioethics to give the communitarian dimension of moral thought greater consideration than previously. The shift toward a communitarian approach is not simply the choice of an alternative method. Rather, the communitarian turn recaptures essential elements of ethical thought that were often misplaced in the Enlightenment project and modernity.

Ethics is a reflection on the moral practices of human beings. This starting point does not exclude animals or environmental questions. However, ethics begins by reflecting on the moral practices of human beings, and ethical questions are asked and reflected upon by particular human beings in specific roles with different obligations and responsibilities (as a parent, patient, physician, nurse, professor, lawyer).

In a sense this starting point seems obvious. But the obvious was often overlooked as modern moral philosophy sought to prescind from the particular and focus on the universal. And, in fact, this "project" has been very influential on bioethics. It is also the reason for philosophy's dependence on the category of "personhood." Persons are a universal category abstracted from particularity. Immanuel Kant, one of the most influential figures in modern philosophy, attempted to put moral philosophy on the fixed foundation of the person, or transcendent subject, and his discovery therein of immutable moral law. Rawls's choosers, behind the veil of ignorance, are a contemporary version of Kant's assumptions. Utilitarianism seeks to move beyond particularism by determining the greatest good for the greatest number. Bioethics, influenced by modern philosophical ethics, has been drawn to general and universal approaches. For example, many questions in the field are formed in the language of personhood: the personhood of the fetus in the abortion debate or the status of patients at the end of their lives.

The conceptual problem for bioethics is that issues in the field are not about abstract persons. Moral practices and decisions are concerned with concrete, particular human beings. We exist as human beings: limited, embodied, particular. That is, we exist as embodied beings and not as abstract persons. In acting as human beings we cannot shed our embodiment or our particularity. Embodiment locates us in the world at a particular place and time and with others.[3] Even the concept of "person" rests on humanity. It is derived from the lived experience of human beings. The concrete, particular exigencies and circumstances of human life shape our moral choices and our critical reflection on them—especially in the fields of medicine and health care where the issues turn on human embodiment.

If bioethicists want to examine how women and men ought to act as human beings, they must begin to look for answers and responses that do not exclude our embodied particularity. Embodiment places human beings in relationships of need and mutual support. It places them in community. The turn toward particularity need not be a turn toward isolation. Rather, it is a turn toward seeing ethical issues within the communal context of particular human beings. Human beings need to be cared for, fed, clothed, and housed. Bioethics in particular deals with men and women in their embodiment and brokenness.

The turn toward community is a turn toward a more complete and complex account of what it is to be human. Just as embodiment particularizes individuals, so do communities have particular understandings of themselves and their history. The task for this chapter is to understand how different moral communities see themselves. The task for the next two chapters is to understand how communities can form societies, and why we are not left with moral relativism.

———— The Shapes of Moral Communities ————

One way to begin an analysis of communitarian bioethics is to examine more carefully the three elements that help define moral communities: content, authority, and engagement with others.

MORAL CONTENT

Ezekiel Emanuel argues that moral communities are shaped by a common notion of a good life: what it is and how it is to be achieved. In a secular, morally pluralistic society one finds a variety of views about the good life and with these views, a variety of moral communities. For example, the good life can mean different things to native New Yorkers, Texas ranchers, and Florida retirees, or to Southern Baptists, Roman Catholics, and Orthodox Jews. These views about the good life also yield different views of bioethics, since they involve particular conceptions of a good death and the meaning of life, suffering, and sexuality, among other things. That there would be a common communal bioethics in a morally diverse society seems to be a failure either to understand the heterogeneity of communities or to presuppose a weak notion of community.[4]

Communities can be bound racially, ethnically, ideologically, or teleologically. Furthermore, the purposes which define a community can be narrow or broad. A moral community may be a group formed to address a particular moral concern such as groups formed to fight pornography or abortion. At the same time, there are all-encompassing moral communities that embody a way of life. Though by no means exclusively, religious communities are clear examples of such all-encompassing communities since they often provide a framework for moral commitments within deep and passionately embraced accounts of the meaning of human life.

The moral formation of members of religious communities involves a total incorporation into the community's interpretative structure (i.e., into its understanding of sexuality, birth, suffering, sickness, and death). All communities have some organizing vision about the meaning of life and how one ought to conduct a good life. Religious communities lend themselves to such use because they frequently have a robust and thick ethic embedded in an understanding of the meaning of life. A religious community, for example, will have a faith that enables believers to understand events of suffering, illness, and death as part of a narrative that speaks to the meaning of life. And, in many of these communities, a link is forged between the moral life and the fundamental narrative that unites the community. One might also think of communities such as the ancient Athenians whose views about human conduct were situated in the context of the meaning of life. People were schooled not only in the meaning but also in how one lived this view of life. Situating the moral life within these broader narratives can give standards for judging the coherence and wholeness to a moral worldview that is often lacking in philosophical accounts.

Moral communities often articulate their views in different ways. They may be articulated in terms of values (e.g., liberty, equality), principles (e.g., justice, nonmaleficence, never take an innocent human life), rules, or the moral life can be articulated in the lives of moral exemplars (e.g., Florence Nightingale for nursing). Engelhardt writes that "[i]t is within particular moral communities that one lives and finds full meaning in life and concrete moral direction. It is within particular moral communities that one possesses a content-full bioethics."[5] Within such communities one learns which moral and nonmoral goods ought to be pursued. One also learns the morally appropriate and inappropriate ways in which goods can be pursued. One learns what promises should be made and the proper attitude toward death and suffering. Communities often teach morality through example. Role models of the virtues are important to the life of the community. Within a community, Engelhardt argues, morality is not simply an aesthetic choice. Rather, it is a truth to be lived.[6]

A community not only has a set of moral commitments; it also has a framework that puts them in relationship to one another. In this setting, members of the community know the gravity of different choices. A Roman Catholic involved in the procurement of an abortion knows that sin is involved.[7] A Roman Catholic who publicly contends that the church is wrong, that abortion is not a sin, commits an act so wrong as to be subject to excommunication. The community has beliefs about matters of proper behavior as well as regarding the boundaries of the community.

A community may also have basic authoritative texts to articulate and shape its moral vision. Yet the community may reinterpret the texts in each generation and circumstance. In this way, different moral communities can share the same texts. For example, Christians in the twenty-first century follow the early church's canon; and Christians, Jews, and Islamic groups put their own gloss on common scriptures. Within each group, differences over interpretation usually also involve different opinions about who has the community's authority to interpret texts. Indeed, one of the dividing issues of the Protestant Reformation involved this issue.

Different moral communities can have overlapping elements of content. The phenomenon of overlapping views is the focus of the second part of this chapter. At this point, however, it is important to note that different communities may hold the same moral values but rank them differently. These different rankings will lead to somewhat different moral worlds. There may also be overlapping principles or paradigm cases, but with important differences in how these principles or cases are understood. If we can understand these differences, we can also map out the terrain of moral acquaintances.

To understand a community's moral view one must also understand its view of moral authority. The function of authority—whether invested in the individual member or in particular offices—is to interpret the moral vision in the light of experience and experience in the light of moral vision. Moral authority is not separate from the vision but integral to it. For example, in a religious vision the authority to interpret the meaning of moral vision and experience is tied to one's view of divine authority. Or, in the liberal tradition, the moral authority of the individual is tied to the underlying vision of the individual and the limits of others. This is particularly the focus of the liberal tradition's view of the limited moral authority of the state. The understanding of moral authority is embedded within the broader moral vision. One type of question where the content of the moral vision and the role of moral authority often intersect depends on whether or not a community understands its moral vision as developmental and progressive or fixed and static.

AUTHORITY WITHIN A COMMUNITY

Communities are shaped by a vision of whom they are. That vision, however, is always open to interpretation and development. The interplay between the vision's content and the meaning and role of moral authority can vary significantly among communities. The extent and nature of moral authority depend in large measure on the extent of the vision. One can imagine a moral community bound together around a particular issue (e.g., the peace movement, abortion, the abolition of communism or genocide, etc.). In such communities, the role of moral authority is often informal and charismatic. Other moral communities may have a thick, robust content set in a larger view about the meaning of life. Here, again, religious communities can serve as a clear example, as each community has moral authorities that help interpret its vision.

Moral communities vary in their understanding of moral authority. Many communities vest authority in "experts" or "moral authorities," such as theologians. Communities may also have men and women who serve as exemplars. These are wise women or men, holy people, saints, or theologians. They are people who know the community's tradition or embody it in some way. A community may also have a juridical notion of authority; that is, it vests authority *in* certain people (e.g., bishops, rabbis, ministers, confessors). Authority rests in one's office, not in the person. At times, these two types of authority may overlap. Moral authority in a community can be tied to knowledge, example, office, or to all three.

Bioethical considerations are not simply isolated principles or propositions but part of the actual life of real communities. How people respond to bioethical issues may indicate whether they are in or out of a community and whether they have a communal identity. For communities of robust, thick commitments and strict observances, the strongest possible penalty against transgressors is excommunication. Religious communities have deployed such penalties (e.g., excommunication, shunning) extensively. Even secular "religions," such as classic Marxism or the Greens, expel heretical community members. Forms of correction, even excommunication procedures, are, as procedures for admission to a community, part of the group's ability to maintain its self-identity. In reminding the community of the offense, these procedures generally reinforce its gravity and justify the harshness of the penalty.

COMMUNITY, SOCIETY, AND CULTURAL ENGAGEMENT

Moral communities in secular societies find themselves living among others who often hold conflicting moral views. Being surrounded by communities with different moral views raises an important question as to how a community can respond to other views and still maintain its moral vision. Each community will, of course, respond according to its own moral narrative and views.

In a postmodern, morally pluralistic world, there is no canonical way to determine how a community ought to respond to moral pluralism beyond the notion that one may not use moral agents without their consent. Here bioethics is, once again, instructive. Secular moral controversies, such as the sale of RU-486 (i.e., the "morning after" pill, an early abortifacient) or physician-assisted suicide, are understood differently by different moral communities. Those who recognize the wrongness of such endeavors confront a spectrum of possible responses.

At one extreme is the so-called "Amish" option or withdrawal from the world. This option does not always, however, represent a "clean" break with the secular world, since some continued interaction is inevitable. The Amish option also poses important questions for the larger society. How can the larger society justifiably act toward communities that are radically different from the general society?

At the other end of the spectrum of communal responses is the "Unitarian Option." Rather than strive to maintain its moral integrity at all costs, this moral community decides to engage the world *in terms the world will understand*. This option, too, involves a significant risk. If the search for agreement with those outside the community is too intense, the searching community can lose its own content. But the community cannot yield its own moral commitment to avoid offending others or to accommodate the character of the age. Communities that seek to engage in substantive moral compromise with moral strangers expose themselves to this risk. The drive to be ecumenical, to bridge differences and particularities, can suppress differences and devour content.[8] Thus does the Ecumenical Content Monster (ECM) threaten to destroy the particular content of any community in search of common moral ground and agreement with those outside it. In this way, some Christian communities have, for example, accepted abortion, assisted suicide, or mindless vitalism.

Between the extremes of separation and capitulation is another approach to moral ecumenism. Fully aware of the ECM, a community can, nevertheless, seek to understand what binds it to others by first understanding how it differs from others. In this model of moral ecumenism, one seeks to understand the core meanings of a community and to identify what it can and cannot compromise for the sake of dialogue. In this way, common ground can be not only identified but also strengthened. This model of moral ecumenism proceeds by way of its differences to the discovery of common ground. It offers a solid foundation for secular bioethics—it neither suppresses difference nor seeks to accentuate it.

The ECM is a particular danger for moral communities who want to address moral problems in the world in terms easy to understand.[9] With hope for common understanding driving us, we tend to smooth over differences. The difficulty is to discover how to address such problems in common with those who do not share the same moral point of view or the same interpretation of human experience. To begin, members of a particular community must recognize their own biases or background assumptions (e.g., the "sacredness of life," or the "common good") and those of others. In an effort to reach those outside the boundaries of a particular community, the community may try to express its own moral vision through the moral languages and concerns of others they confront. In the process moral meaning may be attenuated and the guidance of traditions lost. The result can be that the moral community blends without border into the larger society and loses its particular moral commitments and its integrity. Such losses can occur in what appears to be an innocent moral dialogue, which simply forgoes strong reminders of moral differences and the need to condemn values that must be tolerated. For example, that one ought not commit abortions because human life is created by God and is "sacred" can be difficult to justify to those who do not share theistic assumptions. Similarly, it is difficult to justify, in purely secular terms, that even in the face of inevitable suffering terminally ill patients ought not take their own lives.

Communities can often fail to understand that their most basic assumptions are not rationally self-evident to those outside the community. When this happens, they can become the victim of the ECM. The Roman Catholic response to abortion in the United States, since *Roe v. Wade,* is illustrative of the unitarian danger for moral communities. Holding a long-standing commitment to a natural law tradition of moral reason, the church has engaged, with great vigor, in the public debate on this issue. However, its appeal to "reason alone" has failed to settle the issue in the bioethics arena. As in any argument, the conclusions drawn about abortion depend on the premises with which the argument begins. The debates are not only about the contents of rational propositions but also about the nature of reason and moral justification. On this important topic, feminist bioethics can be most instructive. Feminist writers have raised important questions about how moral justification is conceptualized and discussed. They have also sought to include other forms of moral discourse (e.g., narrative, literature).

The danger for a moral community, such as Roman Catholicism, is that its members may become skeptical about the community's position when rational arguments fail to persuade others. They may question the validity of the community's moral position because the position was advanced not as a religious belief, but as a consequence of rational argument. A failure of reason then engenders a failure of faith since that faith was cast exclusively in rational terms. By asking too much of reason, and too little of faith, faith is brought into question when reason cannot accomplish what can only be accomplished by faith.[10]

What is often overlooked by the church, and other communities, is that for people to reach secular conclusions they must begin with shared moral premises. It neglects conversion in favor of argument. Before reason can work to resolve the content of moral dilemmas, women and men must be converted to the same moral faith. As Stanley Hauerwas observes, for people to agree with the church, the church must first be itself and present a witness in the world.[11] Only when the witness is accepted can men and women reason to more common conclusions. One must first understand the importance of a set of premises within the context of a moral faith before moral controversies can be resolved in common.[12] Many recent discussions about legitimating the religious voice in the public square[13] have overlooked the importance of each community being living witnesses to that which the community believes. One can borrow from Hauerwas and argue that the most fundamental voice for a community is that of its lived witness.

Between the two extremes of complete withdrawal and complete surrender lies the possibility of maintaining a community's identity and still engaging in the world. This position is a commitment to find the points we hold in common even as we acknowledge differences. The exploration of differences helps us identify and better understand what we actually hold in common. The risk of not exploring the differences is that a facile, perhaps fragile, moral ecumenism may flourish for a time. Instead, the focus must be on renewing the identity of the community in the midst of other communities. Orthodox Jews, who have successfully maintained their faith and identity over centuries in a world that is often hostile to them, offer a model by which to imagine this option. They have maintained their identity while engaging in the world in a way that the Amish do not. Though they are engaged in health care and other institutional projects in the larger society, the focus of their institutions is primarily on maintaining the community. Their practice of bioethics is shaped by their tradition.[14]

How to relate to moral cultures outside a particular community is implicitly a question of how to maintain the identity of the community. The interactions with other communities may in fact help a community be more faithful to its own vision of the world and the good life. Other communities may pose, for example, a challenge to prejudices within the life of a community that are inconsistent with the community's professed moral vision. All substantive moral discourse begins in an act of the will, an act of choice about the moral life. Such choices are not always explicit. As one explains and justifies moral choices, basic presuppositions about the moral world become clear. Whether one appeals to the affections or the intellect, an appeal requires moral content. No matter the content (e.g., a belief in maximizing pleasure or a belief in the sanctity of human life), the initial premises of a content-full moral view rely on choice. To have moral identity and integrity while still engaging the surrounding society requires clear awareness of what one can share with moral friends and acquaintances and what cannot be shared with moral strangers. Controversies in bioethics thus disclose the differences that separate moral communities while providing occasions for the differences to disappear.

——— FRIENDS, STRANGERS, AND ACQUAINTANCES ———

The three variables of content, authority, and engagement are helpful tools in analyzing and understanding particular moral communities. They help us understand how communities can share overlapping moral commitments, and they enable bioethics to move beyond particular communities to analyze large-scale secular societies. That is, each of the variables helps us explain how men and women from diverse communities can find common points of agreement on the one hand, and have great disagreements on the other. This insight enables us to examine common morality from another perspective.

Borrowing from H. T. Engelhardt, Jr., I want to argue that men and women can be moral friends insofar as they share a common view of the good life. Members of a particular moral community are surely moral friends; however, in a secular society with different moral communities, there may be moral friends, but many men and women will be moral strangers. The latter are people from different moral communities who have different moral commitments or different rankings of the same commitments. The categories of moral friends and strangers are, however, inadequate for bioethics, which, as a field, has fostered a number of successful encounters as well as experienced some division and difference. To understand its successes one must give an account of how men and women from different moral communities can work together on moral projects. I will expand on Engelhardt's language and argue in the following chapters that a third category, that of moral acquaintanceship, helps explain pluralism in the bioethics community.

MORAL FRIENDS

Moral friends live within moral communities and speak the same moral language. Moral friends share a moral narrative and commitments. This sharing is a necessary condition for moral friendship. Moral friends share common understandings of the foundations of morality, moral reason, and justification. It is moral friendship that grounds the moral community and moral community that supports moral friendship. The stronger the friendship based on common moral traditions, practices, and vision of the good life, the stronger the community. Moral friends share more than abstract sets of principles, rules, or virtues.

Indeed, in such familiar contexts, moral judgments may need little, if any, explanation to those who are moral friends. Friends are able to discover real solutions to moral controversies since they recognize the same moral argument and its conclusions as sound. Moral friendship, however, does not preclude disagreement in that friends may agree on a range of possible actions but disagree on a particular choice. When there are disputes, they know the authority to which they can appeal to resolve their disputes. While the secular tier may be constrained in its vision, its constraints do not preclude the possibility of a rich, content-full moral life within particular communities. In the face of the emptiness of general, secular moral discourse, it is in the discourse between moral friends that one encounters content-full discussions of issues in bioethics.

MORAL STRANGERS

Some bioethicists have focused on questions of morality, public policy, and the interaction of "moral strangers" in a morally pluralistic society. Of course, one of the best known explorers of this area is Engelhardt, who argues that in a secular and morally pluralistic culture, the cardinal moral principle is the principle of permission.[15] Engelhardt also contends that there are two domains of discourse in the moral life. The first (as we have seen) is the domain of moral friends; the second is the domain of moral strangers. This domain of discourse depends on the language of permission and consent.

There are two difficulties with Engelhardt's category of moral strangers. Paradoxically, it is simultaneously too strong and not strong enough. One difficulty is that it is overly restrictive. It too easily separates people as moral strangers. In the second edition of *The Foundations,* Engelhardt says that moral strangers are any who have "differences in moral and/or metaphysical commitments."[16] But this view too easily separates people. Engelhardt fails to take into account the possibility of a moral community that defines itself as open and eclectic. There were, for example, significant metaphysical differences between the Dominicans and Jesuits in the sixteenth-century disputes over grace and freedom, yet it stretches the category to think of them as moral strangers. The classification of moral strangers overlooks many people's experience that others are neither moral friends nor moral strangers. The geography of relationships can be clarified by introducing the category of moral acquaintances. Another way to describe this problem is to say that Engelhardt's category of moral friendship is too narrowly constructed or that he has a particularly restrictive view of friendship.[17] Anyone not in the category of moral friends becomes a "moral stranger" in his account.

At some times, however, the category of moral strangers is not strong enough. While it too easily allows simple differences to separate men and women, it also fails to account for great differences between them. People not only hold different views, but they often hold views that are in deep opposition to each other. They are *moral enemies.* Clearly, other categories are needed.[18]

MORAL ACQUAINTANCES (A₁ AND A₂)

Engelhardt uses the encounter of Captain Cook and the Hawaiians as an example of moral strangers. However, many people in health care do not encounter one another in the way Cook and the Hawaiians did. People often know enough about the moral views of others to understand them. With one type of acquaintanceship (A_1), men and women may understand the moral views of others even though they do not share them. They may, for example, share a commitment to the values of liberty and equality and yet rank them in a different order. Such moral acquaintances are able to have discussions, which are not usually possible for moral strangers whose views are opaque. There is another type of acquaintanceship (A_2) in which the parties involved understand one another's moral world and share it in part. This type of acquaintanceship leads to the very strong claims about moral agreement and overlapping consensus that one finds in bioethics. The accounts of bioethics offered by Beauchamp and Childress or Jonsen and Toulmin are plausible to many because some level of moral acquaintanceship exists.

Many in bioethics assume that something like acquaintanceship (A_2) is at work in secular bioethics. While acquaintances are not moral friends, they seemingly share enough in common that they should not be called moral strangers in any strong sense. For example, one might find utilitarian ethicists and deontologists with basic Christian moral sentiments who have enough language in common to speak with one another and to understand the explanations given by others for the judgments that are made—even though they may not accept these judgments. The view in bioethics has been that acquaintances share enough in common to resolve moral controversies.

One can identify many methods in bioethics that rely on some assumption of acquaintanceship. These methods understand bioethics as resolving moral dilemmas (rather than controversies) and often rely on some appeal to common morality. Yet there is no argument given as to why one should think in terms of a common morality and no clear development of what that morality is.[19] Indeed, as one reflects on different health care issues (e.g., abortion, euthanasia, justice) and surveys the diversity of moral opinion, it is far from clear that there is a common morality.

This appeal is more difficult than it first appears. It misses the difficulties of moral language in which the same word can have many different specifications. To fix meaning is to create a rule, and rules reflect a way of life and a particular point of view. If a meaning is given a high level of abstraction, it can take on such general meaning that any inference can be made from it.

While it is possible to achieve some common understanding, even agreement at times, it goes too far to say that such eclectic borrowing constitutes a common moral framework. The difficulty with the discourse of acquaintanceship is the familiarity of acquaintanceship. Acquaintances share words in common (like "autonomy"), which often leads to an assumption that there is a deeply overlapping consensus. But such conclusions are often too strong. In bioethics, the term autonomy is a good example of such superficial overlapping. Some patients, physicians, and courts may hold that autonomy (i.e., patient choice) is sufficient to justify assisted suicide. Others may understand the realm of autonomy in health care as more restrictive.

Absent some common moral framework, people will lack the ability to determine appropriate descriptions of cases or definitions and applications of principles. The "agreement" of moral acquaintances will often lack a common underlying foundation on which generalizations can be made. Moral acquaintances know that they are not friends, for in understanding the other they also understand themselves and the differences that separate them. Bioethics has too often tended to portray acquaintances as moral friends. This tendency, however, is the equivalent of a misplaced moral ecumenism. The ecumenical movement is on dangerous ground if it avoids talking about significant, deep differences on issues and focuses only on what is, or seems to be, shared. Over time, real differences, if they exist, are bound to emerge. The recent disputes on the ordination of women, for example, highlight deep differences over the "development" of tradition, the nature of priesthood, authority, and ecclesiology.

—— Summary ——

The turn toward moral community is a return driven by a view of ethics as practical knowledge tied to moral practices within a way of life. A communitarian perspective is concerned with how human beings should act qua human beings in their embodied particularity. However, moral communities are complex and there is no single model to capture them all. There are at least three variables that shape moral communities: content, authority, and engagement with other communities in the public square. In spite of strong differences there is no reason, a priori, to hold that communities cannot overlap. A consideration of methodology in bioethics needs to address how to account for this overlap. Health care, in secular societies, is a collaborative enterprise and moral problems are not contained only within the boundaries of particular communities. Nor do most men and women live, strictly, within the boundaries of a particular moral community. The categories of moral ecumenism and moral acquaintanceship provide a way to understand and map the different intersections that can and do take place.

The turn toward community enables us to identify certain elements that are part of ordinary moral discourse. It also enables us to identify the complex range of relationships men and women can have as moral friends, acquaintances, and strangers. While the conceptions of the good life vary along the spectrum, the overlapping nature of moral acquaintanceship explains why different methodologies work or do not work in bioethics.

6. The Genealogy of Agreement and Consensus

From the clinic to medical research to health policy, bioethicists seek agreement or consensus in some form about difficult cases and issues that arise at crucial moments in human life: conception, birth, suffering, and death.[1] Bioethics has cast itself as a discipline that resolves moral controversies in medical research, experimentation, clinical treatment, and health care policy. As a field of inquiry seeking to resolve moral controversies, bioethics has sought agreement or consensus with a zeal reminiscent of the knights' search for the lost chalice. Each method in bioethics attempts to establish as much agreement as possible, and different methods legitimate themselves, in part, by their ability to articulate agreement. Indeed, a criticism of some methods is their inability to resolve moral dilemmas. Even the meaning of a "successful" consensus can be interpreted in different ways. Beauchamp and Childress have described agreements of the National Commission as agreements about principles while Jonsen and Toulmin describe them as agreements about cases.

How much agreement is necessary for bioethics? And what actually is being agreed to? The answer to these questions depends on one's methods. Beauchamp and Childress, for example, affirm a common morality and four middle-level principles as points of agreement. Specification of the principles, however, permits a wide spectrum of difference. Jonsen and Toulmin also assume a common morality but look for agreement in cases. Some agreement must be reached to determine the paradigm cases and how they ought to be interpreted. Even Engelhardt, who puts much greater emphasis on disagreement and controversies, speaks of fundamental agreement at the level of permission. Jonathan Moreno, in his work on consensus in bioethics, argues that a consensus has formed around certain issues.[2] Each method argues for a type of transcendental agreement. That is, each method understands its starting point—principles, cases, permission—as the necessary condition for secular bioethics. The argument of this chapter is that although agreement has been overrated by some (e.g., Moreno) and underappreciated by others (e.g., Engelhardt), exploring agreement and consensus can lead us to an understanding of moral acquaintanceship.

Three questions can be raised about appeals to agreement or consensus in bioethics. First, we must determine where to locate the agreement; is it in cases, principles, or consent? Different methods in bioethics claim different points of agreement. Second, how do the agreements cited by different bioethical methods enable us to understand our disagreements? Though we speak about agreement and consensus, disagreement is often the unspoken text of bioethics. Nicholas Rescher has used the term "dissensus" to contrast areas of disagreements with agreement and consensus.[3] Third, what is the moral justification for our agreements or consensus? That is, why should consensus be taken as morally normative. The histories of race and gender relations are filled with examples of moral norms that were accepted by the status quo and had consensus support that were later rejected on moral grounds.

One response to these questions is to ask whether we do really agree. Disagreement in bioethics is most striking in the ongoing controversies over abortion or physician-assisted suicide. Yet some bioethicists, for example, Moreno, are adamant that we can achieve a "bioethical consensus." On analysis, however, Moreno's consensus is a consensus about procedural ethics and consent. Even so, more analysis is needed. At this point, however, it is helpful to keep in mind that those who participate in a consensus or agreement often have very different understandings of their achievement. Another response to reported agreements is to argue that they were reached because they are reconstructions of the same moral world. They result because those involved have similar moral views and more in common than they may have realized.

Neither of these responses adequately captures agreements and disagreements in bioethics. The turn toward community in the last chapter and the category of moral acquaintanceship offer a more complete way to understand epiphanies of agreement and disagreement in bioethics and the dissenting opinions that accompany them. As we noted in chapter 5, at least three central elements, in one form or other, shape a community's moral vision: content, authority, and engagement with others. Each of these elements represents an opportunity for agreement or disagreement with others—especially engagement with those outside the community. It is possible to have agreement or disagreement at any of these levels. For example, moral friends may agree strongly on content; moral strangers may be satisfied with procedural agreements; and moral acquaintances may develop limited, overlapping, substantive, and procedural agreements. Agreement can be found in many different places and in many different forms. One might think of agreement as similar to U.S. Supreme Court decisions. A decision can be unanimous but still admit subtle and important differences that the justices will express in their opinions. For example, the Court has unanimously upheld the distinction between killing and allowing to die. Still the wide variety of reasons and arguments articulated in the justices' written opinions shows how fragile the agreement can be. Chief Justice Rehnquist frames the constitutional issue in terms of the role of state legislatures, while Justice Souter frames it in terms of ordered liberty. Justice O'Connor agrees with them on upholding the statutes but writes on issues the Court did not address. If the cases were different, one wonders if the agreement would hold.

Thus the agreements claimed by bioethicists as evidence of a common morality are always in need of careful analysis. Sometimes the cited agreement reflects consensus about a particular action; sometimes it reflects consensus about principles or values. At still other times, the agreement may be accidental or simply reflect the outcome of a managed process. Imagine, for example, Peter Singer and Cardinal Ratzinger as members of a bioethics commission discussing the treatment of very low weight neonates. They may agree to limiting treatment for deformed newborns, yet they will reach this decision on very different grounds. Singer may argue that prolonging a life of the premature 500-gram neonate needlessly prolongs the misery of the newborn and its family.[4] He would stop all treatments to maximize utility. Cardinal Ratzinger might well agree that treatment should be withdrawn, but only because, while we have a basic obligation to conserve human life, that obligation is not absolute and one can licitly withdraw treatment when there is little hope of recovery. The lack of a univocal argument becomes even clearer if the case is modified. If Singer argues that one should actively euthanize the newborn, Cardinal Ratzinger will sharply disagree, and the disagreements become more pronounced with each added nuance. One begins to see that agreement occurs primarily in the extreme or easy cases, and only by choosing the easy cases do the claims of agreement actually work.

Family resemblances among moral vocabularies are often mistaken to represent sameness rather than resemblance. This mistake, in turn, contributes to the confusion about agreement. When several things share a resemblance they are not the same. Rather they resemble one another in the way membership in a family does. There is an overlapping, "crisscrossing" of features.[5] Three sisters, Mary, Carolyn, and Mary Rita resemble one another. Mary has Carolyn's eyes while Mary Rita has her nose, and Carolyn has Mary's hair. There is a similarity but no common feature or essence. Those who place emphasis on agreement to avoid the dilemma of endless foundational disputes fail to recognize that the similarities in middle-level principles or between cases is a family resemblance and not the disclosure of a common underlying feature or essence. The resemblances are real but there is no one underlying isomorphism binding them together. Family resemblance may reveal a type of moral acquaintanceship (see chapter 5). That is, men and women are morally acquainted when they share enough moral vocabulary and values in common to allow them to understand one another.

—— Pluralism and Consensus ——

Consensus can take place at a number of different levels: at the level of belief, it affects theory and cognition; at the level of action, it is pragmatic and practical; and at the level of values, it enables coherence and motivation. For consensus to play an important role in bioethical method one needs to understand which of these levels is being asserted. Thus, it becomes important to ask why a consensus exists.[6] Is it mindless conformity? Is it about a submission to or support of existing power structures? Is the consensus driven by the weight of appropriate evidence? Nicholas Rescher suggests that one should ask whether the consensus being appealed to is an idealized version of consensus or one that is practically attainable. Philosophers tend to use the former while social scientists deploy the latter.

Rescher notes that any talk or use of consensus must also investigate dissensus. Consensus and dissensus, like health and disease, are dialectical terms, and one cannot be understood without the other. That there should be dissensus in bioethics is not surprising. If morality is part of a way of life, and ethical reflection is grounded in moral experience, then different experiences will lead to different views. Of course one response to the different views and methods captured under the phrase "pluralism" is some type of relativism or indifferentism. The relativist view is that it really does not matter which position one holds on any matter. However, a problem with this view is that if one holds it, he or she will have no incentive to reach a consensus with anyone who holds different views. There is no reason for anyone to negotiate a consensus if he or she has no reasons to hold any position whatsoever. The relativist view also leaves us with no intellectual or moral argument against the use of power to impose a position.

Another response to pluralism is syncretism: a desire to bring together the sum total of all perspectives.[7] But syncretism, like relativism, provides no grounds, in principle, to discriminate among different views. Presumably one's positions about bioethics or particular bioethical issues are based on a set of reasons or moral commitments, and presumably those who make such commitments think that their commitments are superior to others—but syncretism does not help them make this judgment.

An alternative approach, articulated by Rescher and helpful for bioethics, is "perspectival pluralism."[8] This position holds that one needs to have the "courage of one's convictions." One needs to know the positions she or he holds and how they differ from other positions. Such knowledge is crucial to compromise and consensus.

The nature of agreement, disagreement, consensus, and dissensus is best understood within the context of moral judgments. Moral judgments should be understood not simply as choices about what should be done in a particular situation, but as involving logically prior judgments about how one justifies such choices. One's assumptions about moral rationality are prior judgments that commit one to a particular view of the moral world. For example, those in the natural law tradition understand moral rationality in a different way from those who deploy an instrumentalist view. Charting the geography of judgment reveals a number of points for potential agreement and disagreement for moral acquaintances. This mapping of judgment is the task of the second section of this chapter. The mapping process can broaden our understanding of agreement and consensus. The fourth section of the chapter builds on the geography of judgment to understand further the relationship of moral friends, strangers, and acquaintances.

Finally, the question may be asked: How much agreement is needed in bioethics? The answer depends on what is being sought. People may reach agreement to discontinue aggressive life-sustaining treatments of a patient for different reasons. One person does so to allow a patient to die, while the other does so to increase the patient's comfort. Agreement is reached on what to do, but not on why it is to be done, and the agreement is limited to the case. It may be impossible for these agents to generalize from the particular agreement to end treatment to agreement about institutional policy, law, or public policy. This question about how much and how far agreement is needed in bioethics reflects a deeper question: What is bioethics? The nature, quantity, and extent of agreement required in particular cases will depend on how the field is defined. When bioethicists boast of different types of agreement, they are really offering different views about what the field of bioethics is. Indeed, one has to ask if there is a field of bioethics and not merely a collection of loosely related enterprises.

―― The Geography of Moral Judgment: ――
A Spectrum of Agreement and Disagreement

The geography of moral judgment is complex, and knowing this geography is necessary for understanding moral agreement and consensus. Nevertheless, it is often overlooked. Particular moral judgments contain a number of logically prior judgments about moral reason, moral justification, and moral values. Failure to articulate how these different judgments contribute to concrete moral judgments lies behind the mirage of agreement.

The first level of judgment is "object-level" judgments. Object-level judgments concern what one ought or ought not to do in a given case and the particular course of action that should be followed. Casuists are generally concerned with object-level judgments. In our earlier example, Cardinal Ratzinger and Peter Singer were making object-level judgments about the care of a low birth weight neonate in their role as members of a functioning bioethics committee.

Casuists often point to the "agreement" between people like Ratzinger and Singer to argue that we need only be concerned with this level of judgment. However, we must not be swayed into thinking that this level of agreement is more powerful than it really is. Just as "facts" are noted and understood only within a conceptual scheme, so too "cases" are recognized only within a moral scheme. Similar judgments made at the object level, though they may enjoin the same course of action, often represent very different judgments. Like the Roman numeral "I" and the English pronoun "I," these judgments look the same, but they are rooted in very different symbolic systems. The difference between them becomes apparent as one moves from the coincidence of judgment in one case to other cases. Cardinal Ratzinger and Peter Singer share a similar judgment about a particular newborn. However, they do not share enough in common to agree on other cases.

To name and resolve other cases and build paradigms, we must make other judgments. We must, in fact, move to a second order of judgments. Judgments about cases made in the light of logically prior judgments are "justificatory" judgments. These judgments articulate the reasons, moral feelings, moral sense, or moral values that apply to particular moral controversies or dilemmas. The principlism method is an attempt to build agreement at the justificatory level.

Finally, at a third level of judgment are choices that are logically prior to both object-level and justificatory judgments. These choices are about moral foundations; they define the basis and acceptability of justification (i.e., its underlying reason, values, intuitions, sense, and affections). In fact, these foundational judgments control the meaning of the justificatory judgments and their relationship to one another.

The difficulty with the agreement of the casuists—that is, with object-level agreement—is that what has been agreed to is not always apparent. Agreement is not univocal. This difficulty becomes evident when one tries either to describe such an agreement or to generalize it. For example, people with a variety of moral views may agree in general that medical treatment, including feeding and hydration, should be withdrawn from a patient in a permanent vegetative state (PVS). Agreement on the course of action does not, however, imply agreement on the level of justification. Some may want to stop aggressive medical treatment to ease the burden of treatment while others may want to because they believe the quality of life in a PVS patient is not worth preserving. Those who act for the first set of reasons may oppose assisted suicide; those who act for the second set of reasons may support it. One must exercise caution so that a mere accidental coincidence of opinion is not mistaken for agreement. The potential for disagreement is realized in moving from the particular case to judgments about reasons. We can no longer simply say that "X" ought to be stopped. Instead, we must say that "X" ought to be stopped because it is a case of "Y." The causative phrase is the reason, principle, or maxim that links the old case (Y) and the new one (X). Without agreement on the justificatory level, we cannot move from one case to others, and we have no authoritative way to interpret agreements reached at the object-level.

Absence of agreement on the justificatory level limits our ability to generalize. Our choices about what to do in a case reflect different descriptions and diverse justifications about our choice of action. For some, the withdrawal of treatment causes the patient's death and is morally culpable. For others, the patient's death is an unintended side effect of withdrawing treatment and no culpability attaches. For yet others, it simply follows from the wishes of the patient, and finally, there may be some who do not see any dilemma or controversy in this case. Understanding these differences helps explain the fact that people can reach agreement on a treatment program for a PVS patient but not on physician-assisted suicide or euthanasia.

The trolley-transplant paradox provides an illustrative body of literature on this point.[9] This paradox, discussed in chapter 4, is often cited as an example of secular casuistry. The paradox is that we consider it morally correct to throw the track switch so that a runaway trolley only kills one person and avoids killing five; yet we find it morally opprobrious to kill one person in order to harvest his organs to save the lives of five dying people. In the literature there is general agreement on the object level: one should throw the switch and save the five, but the transplant doctor should not kill the one to save the five. The literature, however, is also filled with a variety of explanations and diverse judgments about what constitutes acceptable reasons for these choices, and the disagreement escalates as the paradox is redescribed and extended to other cases.

Agreement on the level of justification is potentially more powerful than agreement about particular cases at the object level, and such foundational agreement may seem unassailable. It is, after all, about reasons and principles; it allows one to understand a whole set of cases. However, in the midst of moral pluralism it may be difficult to move from the level of agreement to the level of cases. Because of different specifications and configurations, people may arrive at different decisions about cases.[10] We are deceived if we think that the same words or "principles" necessarily carry the same moral content. For principles to have univocal meaning they must reside in a univocal justificatory and explanatory context. It is within the foundational level that one will find the language to explain and justify one's choices. Cardinal Ratzinger can speak, for example, of neonatal intensive care units as an extraordinary means of treatment that places an undue moral burden on the newborn and its family. He will then be able to generalize to cover some cases and exclude others.

Unless people share the same foundational framework, they will not necessarily share a common understanding of how principles or reasons should be interpreted. Not all explanations of futility, extraordinary means, burdens, justice, or beneficence are the same. Moral language is often equivocal, and on the level of justification, different shades of meaning can be given to "agreement."

Imagine a situation in which agreement is complete on the object level but no consensus has been reached about reasons. For example, an Orthodox Catholic and a Roman Catholic agree on the evil of abortion but offer different reasons for their judgment. Some of their reasons may overlap. For example, both would profess an obligation to follow God's command against killing human life. However, the Roman Catholic may also deploy views about the ensoulment of the fetus and our natural obligations to the fetus as a person. This argument may be unnecessary according to the Orthodox Catholic.

The foundational level provides the basis for the most powerful type of agreement. What is shared is a common moral language, values, and assumptions. This common framework allows both agreement and disagreement. Recall, for example, the Roman Catholic debate about craniotomies at the turn of the century; at that time, theologians, sharing the same basic framework, reached different moral judgments and justified them with different reasons embedded within the tradition.[11]

And the converse is also true: people can share the same substantial moral commitments and foundational matrix and still disagree on object-level judgments. Imagine two people sharing the same moral commitments, language, and range of possible responses to a moral controversy. For example, two Roman Catholics will see abortion as a moral evil to be abhorred, avoided, and condemned. They will also share the same basic conceptual framework and language for discussing the issue. However, since part of the framework of moral reasoning in this tradition extolls prudence as the cardinal virtue of practical reason,[12] these Catholics may come to different conclusions about what course of action is the best response to the evil presented. Yet both are firmly grounded in tradition.[13]

Even those sharing the same moral framework can disagree at the level of reasons. Two Christians, for example, might agree that each should donate a kidney for transplantation but disagree about the proper reasons for doing so. As transplantation involves surgery (the mutilation of the body) and the impairment of bodily integrity, it calls for justification. Traditionally Roman Catholicism has justified surgical mutilation by arguing that it offers a good for the whole person.[14] Some have adapted this principle of totality (the good of the whole) to justify transplantation because it serves the good of the social whole. Others, however, have argued that this justification is a misapplication of the principle of totality because it gives too much authority to society and the state. The act of transplantation, they argue, can only be justified by an appeal to the principle of charity.[15]

Another example—this one cited by Jonsen and Toulmin—is the discussion of usury in the Roman Catholic tradition and the different reasons cited in support of different conclusions. In the midst of much disagreement, the disputants shared a basic theoretical model, basic value commitments, and a moral language through which they were able to speak, argue, and express their disagreements.

The reality of moral pluralism in a secular society illustrates that there are many ways in which to construct the categories of the moral world. By distinguishing the three levels or types of judgment (i.e., object, justification, foundation) involved in moral argument, the spectrum for possible moral agreement and disagreement is greatly increased. It ranges from a strong sense of agreement, in which we are of one mind on how and why to proceed, to a weaker sense of proceeding together but only for a specific, limited venture.

The complex spectrum of relationships that lies between complete agreement at the levels of object, reason, and foundation to complete disagreement on those levels can be summarized under eight headings.

1. Object-level agreement with agreement on justification and foundations.
2. Object-level agreement with agreement about justification and disagreement about foundations.
3. Object-level agreement with disagreement about justification.
4. Object-level agreement with agreement/disagreement in part on the levels of justification.
5. Object-level agreement with disagreement about both justification and foundations.
6. Object-level disagreement with agreement on justification and foundations.
7. Object-level disagreement with justificatory agreement/disagreement in part.
8. Object-level disagreement with disagreement about justification and foundations.

The possibilities and the limits of each genus of controversy resolution in bioethics can be analyzed under these eight headings. To reach agreement regarding justification there needs to be prior agreement on what counts as a relevant moral appeal and what is a proper set of moral reasons to which one could turn. Unless moral agents stand within the same foundational framework, they will not reach agreement on how moral judgments are justified.

One finds here a genus of agreement that contains different species of agreement. For example, there is agreement when moral agents share and similarly rank values, moral commitments, and moral frameworks. This kind of agreement characterizes moral friends. The agreement wrought by procedures alone is the agreement characteristic of moral strangers. A more limited agreement can be found in some cases or in the expression of some principles (e.g., in death and dying cases and the appeal to autonomy). The extent of this species of agreement is not always clear; it may be limited to a particular case or situation, and we may not be able to generalize beyond it. Lastly, there is managed agreement, which is reached when people seem to arrive at the same judgments without sharing the same moral framework and ranking of values. This agreement is characteristic of moral acquaintances.

—— A SOCIOLOGY OF AGREEMENT ——

The work of various bioethics commissions and committees provides examples of moral agreement in a secular, morally pluralistic culture. Given that commissions have played an inspirational role in the development of bioethics, it is important to examine how such committees and commissions achieve agreement. The sociology of such commissions raises important and interesting questions about what conclusions can be drawn from their work. The first question bears on the composition of these committees. Usually people who are selected for such work are, at least, moral acquaintances. One rarely finds individuals with strongly different views, like a Pat Buchanan or a Jesse Jackson, appointed to the same committee or commission. In the selection of members, the committee's agreement is already being managed. A second question focuses on the committee's process. Such groups are shaped by a dynamic toward reaching a consensus.[16] The expectation before the committee begins work is that it will reach consensus on certain recommendations. Chapter 5 distinguished between an ecumenical spirit driven by a concern to highlight only agreements and an ecumenical spirit that begins by acknowledging differences as well as commonalities. The former is a "cheap" moral ecumenism that is unwilling or unable to do the hard work of looking at real moral differences. All too often such ecumenism is part of the *modus operandi* of such commissions and committees. Their dynamic highlights points of agreement and plays down disagreements. A third question focuses on the establishment of the agenda of the committee. Insofar as the committee is mandated to act in certain questions (and not in others), the possibility of disagreement is reduced.

Notice how the work of such groups contrasts with the exchanges between individuals with great moral differences. One can imagine Cardinal Ratzinger and Peter Singer agreeing in a chance meeting in the Fiorenza Hospital that the treatment of a PVS patient in the Medical Intensive Care Unit should not be continued. However, as they discuss the mission of the hospital and the practices it should or should not countenance, the limits of their agreements emerge. They do not agree, for example, that Fiorenza Hospital should distribute condoms as part of its care for HIV-infected patients. As individuals are apt to have wide-ranging discussions of issues, their disagreements often become apparent very quickly. Unlike Cardinal Ratzinger and Peter Singer the committee will not, officially, wander from its topic. The agenda of such committees gives direction and limits to their moral reflections.[17]

The control of the agenda is a crucial point often overlooked in the heralding of agreement by committees. A necessary condition for resolving a moral dispute is consensus regarding the essence of the dispute. So often in bioethics the most difficult problem is the lack of a common description of a moral controversy (e.g., abortion, assisted suicide). Is abortion about rights, fetal life, or the killing of an innocent human being? Is physician-assisted suicide an act of mercy or an act of murder? If an agenda is established before a committee or commission begins its work, then the mapping of a general moral geography has already begun. The agenda not only identifies the problem, but also provides a way whereby differences are confined and minimized.

Understanding these sociological elements leads one to be cautious about how one should evaluate the claims of agreement. It is helpful to remember that agreements and disagreements can be found at a number of points in bioethical discussions. We simply need to be clear on what is being agreed to and not make extravagant claims.

Excluding thinkers like Buchanan or Jackson from commissions means that fundamental questions about the agenda or agreement will not be asked. Consider Jackson and Buchanan's participation on a national commission evaluating federal research and the funding of reproductive technologies. Conceivably they may agree that such research should be stopped. Yet they may argue from different reasons. Buchanan may see the research as an immoral tampering with the natural order and an unjustified use of state authority and resources while Jackson may regard such research as unjust because its results cannot be distributed equally to all and because it takes necessary resources from the poor. While they agree on a policy recommendation, few would see this as a hopeful sign for future moral deliberations.

James Childress provides an interesting and instructive case study in the management of agreement and consensus in bioethics. Childress examines the deliberations of the Human Fetal Tissue Transplantation Research Panel (hereafter, HFTTR). In 1988 a moratorium was declared on the use of federal funds for HFTTR by Robert Windom, then assistant secretary for health (U.S. Department of Health and Human Services). The National Institutes of Health (NIH) appointed the HFTTR panel in the fall of 1988 to respond to ten questions raised by Assistant Secretary Windom.

Even before it began work, Assistant Secretary Windom and the NIH had given the HFTTR panel significant help in its task since the framing of issues directs the ways in which any moral problem can be resolved. The framing process itself can make the moral pluralism of a committee more manageable. In the case of the HFTTR panel, Assistant Secretary Windom had set the agenda in his ten questions. Childress notes that Windom's questions focused on the linkage between abortion and HFTTR practices. Indeed, Childress argues that Windom's questions constrained the panel's deliberations.[18] Childress himself makes the point that a different set of questions could have led to different outcomes. What is of interest here is that the process of deliberation—and its outcome—were helped and directed by the charge given to the panel. As one looks to the agreements and consensus of panels, commissions, or hospital ethics committees, one needs to examine how the boundaries and agendas of deliberation were established.

Childress also addresses the issue of dissent in the panel's work. He says that two of the eleven members had substantial dissent. The two dissenting panel members produced a dissenting report, such that "panelists in the majority later expressed their concern that such a long and eloquent dissent would simply smother the report's brief responses."[19] Childress notes that an additional meeting of the panel was called to structure the form of the final report so that it would not be overwhelmed by the dissenting report.

The discussion of dissent raises two important questions. First, how much agreement is necessary for a consensus? If a committee is unanimous, the consensus is obvious. However, absent unanimity, and when there is strong dissent, the degree of consensus is difficult to ascertain. Second, is the consensus based on the moral issues? A consensus report may play on certain ambiguities. Childress, for example, points out that the questions raised by the assistant secretary were empirical, legal, medical, scientific, and moral.[20] As one listens to claims of consensus it is important to determine whether the consensus is actually about the moral questions.

Childress's observations remind us that when people assert agreement, it is important to know what types of questions were asked and agreed to. His account raises anew the question of how and what kinds of agreement are possible in a secular, morally pluralistic society. Contrary to the Jonsen-Toulmin experience, Childress cites agreement on the level of principle.[21] It is possible that different methods of bioethics may be appropriate to different activities. For example, issues of public policy, or institutional policy, may be better articulated as principles insofar as principles give broad guidelines for actions. At the same time, particular clinical issues may be better addressed by the agreement of cases. Since method and content cannot be separated it is clear that different methods reflect different moral views.

Committees and commissions have come to play a central role in bioethics. From local hospitals' and nursing homes' ethics committees to national policy commissions, committees have taken on important roles in moral deliberations. As one examines the work of such groups, one becomes aware, however, of the importance of power and control in guiding the resolutions of such committees. The power to set the agenda, membership, and timetable are crucial to reaching any agreement. The Childress account helps us to understand how the agreement of such commissions is managed. It relies on both the agenda of the commission being set and the members of the commission not dissenting in bad faith. That such agreements are managed should not be surprising. Governments, like the people who run them, often seek the opinions of others to support a desired policy or to suppress an unpopular one. The health care task force of the Clinton Administration assembled an ethics task force. Members of the task force shared some common assumptions about society and health care that were important for their deliberations.[22] It is not hard to imagine how the conclusions of the committee would have been very different had its membership been altered in substantial ways.

Members are selected and agendas are set so that a desired result may be achieved. The members of the commission, unlike the Senate (in its role to advise and consent), are bound to the agendas given them. What emerges from this account is a picture of agreement that is often carefully managed and crafted. The result may be an agreement that is more causally achieved and less rationally justified than we craved. This confusion about the nature of agreement occurs often in bioethics. The tendency is to draw principled conclusions when the conclusions are more sociological in nature.

——— Agreement between Friends, ——— Strangers, and Acquaintances

In some obvious cases and controversies in bioethics, the people involved shared very little or nothing morally in common. The continuing warfare between the pro-life and pro-choice advocates in the abortion debate is an example of such an encounter. The parties do not even share a common language. They meet as moral enemies or, perhaps, as moral strangers, notwithstanding that how the debate is framed and how moral commitments are articulated can lead to different conclusions. Often the debate about abortion is not about the act itself but about the appropriate role of the state (e.g., laws). If issues about government authority were separated from the discussion of abortion itself, it might be possible to identify common ground. One way for moral acquaintances to begin addressing difficult issues, such as abortion, is to untangle questions that are often mixed together. Not many, for example, would hold that abortion, in general, is a good thing. Abortion often embodies a tragic choice in some way or other. Acquaintances could begin from this point of agreement to ask further questions about how to build a society in which abortions are less prevalent. Some will surely hold that the best solution is government intervention; others may sharply disagree. But even those who disagree on the role government should play can help identify other strategies to reduce the prevalence of abortion.

There are other situations in which people meet as moral friends. Such friends can be members of the same moral community, share common moral values, and speak a common moral language. Moral friends may disagree but understand their disagreement within the context of what they share. A more difficult situation conceptually is when people meet as moral acquaintances. They are neither friends nor strangers. They share enough in common to understand one another's positions and, in some circumstances, to reach agreement about a particular case, principle, or policy. Such agreements can obscure underlying disagreements and make the diagnosis of moral pluralism seem trivial. But as we will see one ought not to overplay or underplay the importance of such agreements.

The preceding chapter examined some of the different factors that shape moral communities. In so doing it supported the conclusion that women and men can find themselves as moral friends, strangers, or acquaintances. The geography of moral judgment expands our understanding of the relationships among friends, strangers, and acquaintances. Moral friends can agree at the level of foundations; thus, they provide the clearest example of a consensus ("feeling together"). They are members of moral communities that have common values, assumptions about moral justification, and moral authority. Moral acquaintances may agree coincidentally on the object level and perhaps on the justification for their objective judgments. Acquaintances can share bits and shards of the past or some overlapping moral values. Bioethics has made a great act of faith in these types of "agreements" even though the agreement of acquaintances is much more limited and tenuous than the agreement of moral friends. Acquaintances can share general principles, yet have different interpretations of the meaning, ranking, and scope of application. Hence, the importance of the method of principlism in bioethics. Confusion can result, however, when the agreement of acquaintances is treated as the agreement of friends or when their disagreements lead people to think of them as being more like moral strangers or enemies than acquaintances.

The agreements of moral strangers are generally confined to a principle of permission. It is interesting to note that Jonathan Moreno mixes together procedural consensus with substantive consensus and agreement in bioethics. These types of consensus are different by degree, and the degree of difference is born out as the more substantive, content-full questions are addressed. Nevertheless even the level of procedures (a thin level) requires some underlying agreement or overlapping moral commitments for the procedures to work (see chapter 7).

—— Summary ——

As one examines the interrelationship of the different types of judgments that are integrated into any concrete moral judgment along with the overlapping and fragmentary nature of moral views and language, one comes to recognize that bioethical agreements and disagreements are not always what they seem to be. Agreements are not all the same. Nor are all disagreements the same. The more one understands the complexity of moral judgments, and the various types and degrees of agreement, the more one understands how limited the force of agreements often is and how they often mask important disagreements. Simultaneously, disagreements may not be as stark as they first appear.

The geography of agreement and disagreement involves a complex dialectical relationship, and understanding this complexity helps map an understanding of moral friends, strangers, and acquaintances. The points of agreement-disagreement can relate to a specific topic (e.g., a case or a policy), to the reasons for particular choices, or to one's foundational views. Moral friends will have agreement at the level of justification and foundations. Moral strangers may have disagreement at every level. Moral acquaintances will often have more complex and ambiguous forms of agreement and disagreement.

Any robust or thick sense of agreement can take place to the extent that common moral frameworks and commitments are shared. It is most obvious that moral friends, in sharing the same understanding or moral reason and the same framework of moral values, share a substantive or content-full view of the moral life. In this context, they can ask one another's advice about judgments to be made and reprove or praise one another for judgments that have been made. They can also see, among themselves, how strange and disordered the moral world of others is whom they encounter in general secular society. The depth of disagreements will depend on the extent to which they have something or nothing in common.

Moral strangers can share agreements based on consent and permission. The strength and extent of agreement among moral strangers will vary from agreement to agreement. The agreements of moral acquaintances will be somewhere in between. Such agreements may be around particular issues or goals (e.g., abortion, national health care) or around choices of actions and the reasons for them (e.g., not prolonging the death of a patient).

Some will ask if these concerns about agreement are important. How much agreement does bioethics really need? It seems clear that each of the different methods in bioethics assumes some type of agreement or consensus. However, the constitution of agreement is variously accounted for in each method (e.g., in cases, principles, common morality, virtue, or permission). Such different accounts of agreement underscore the ambiguous nature of agreement in bioethics and the ambiguous nature of bioethics itself. How one defines bioethics and its scope will influence one's views about agreement.

Yet agreement is, nonetheless, important. Even the weakest sense of agreement relies on an overlapping moral consensus or agreement. In bringing together an analysis of agreement and the category of moral acquaintances, the stage is set to look at procedural ethics as a way to discover common ground. One can understand the thin agreements of procedural ethics only if they are built on thicker, richer understandings of the moral life. Absent such overlapping values the procedures could not succeed ethically. Procedures need some form of moral justification if they are to be moral. If there are procedures that transcend moral communities then they may provide a way to identify the common ground of moral acquaintances. The agreement about procedures provides a way to articulate the overlapping agreements that exist for moral strangers and acquaintances. The next chapter argues that without these underlying agreements even a procedural ethic would not be possible.

7. Moral Acquaintances: Proceduralism and Organizational Ethics

Bioethics is a dialectic shaped by ethics and morality. If morality is part of a way of life—a set of practices that judge human conduct as either right or wrong and human beings as praiseworthy or blameworthy—and ethics is the attempt to articulate justifications for moral practices and reflect systematically on morality, then we can say that bioethics seeks to be both a morality and an ethics. As practiced, bioethics is between a first- and second-order discourse (i.e., between morality and ethics) and has features of both. The field addresses concrete, particular moral controversies in medicine and health care: issues in the clinic, the laboratory, and the public square. It often recommends particular choices and condemns other choices. At the same time bioethics attempts to move beyond particular issues to think systematically about the field and search for justifications for particular recommendations. Questions of methodology are essential to both levels of discourse. How one develops ethical justifications sets boundaries on moral practices, dilemmas, and controversies; and assumptions about moral practices direct ethical reflection. It is the dialectic that makes methodological assumptions crucial.

Indeed from its origin in theological and medical traditions, bioethics has been variously deployed. This methodological pluralism is not surprising for at least two reasons. First, the field of bioethics comprises several disciplines. Law, medicine, philosophy, and theology played important roles in its development with each discipline bringing different methodological considerations to the field. Second, secular bioethics entails a pluralism of ethical methodologies, since part of the postmodern condition is the lack of a universal narrative or common moral culture. Methodological choices reflect different ways of living in and viewing the moral world. Note, however, that while different ways of life may lead to different moralities (i.e., pluralism), this fact alone does not necessitate methodological pluralism. For example, both Judaism and Roman Catholicism have similar but different moralities and both have used a model of casuistry. But similarities are not "sameness." Nonetheless, contemporary bioethics has encountered different moralities and different methodologies.

What directions can we take from the pluralism of methodological choices? One direction is to smooth over the differences as if they do not exist. But this solution leads nowhere because the ignored differences will simply reemerge. Another direction is a kind of communal relativism that accentuates the differences. This direction is also not helpful for bioethics. Medicine and health care are not merely private or local enterprises. They are social behaviors practiced in the setting of a secular society that encompasses a variety of moral communities. However, moral pluralism does not necessarily mean that moral communities are separated or isolated from one another. Nor does it leave us with only moral relativism. As I argued in chapter 5, it is possible to speak of a range of moral relationships in secular societies. People can meet as moral friends, strangers, and acquaintances. In chapter 6, through an exploration of the terrain of agreement and consensus we found reasons to carefully scrutinize lines of moral agreement and consensus. The task is to identify, in some way, what the common ground for a secular bioethics might be and how that ground can be used to critique methodological choices. Others have sought, with varying degrees of success, to identify morally common ground through principles, cases, theories, virtues, and the principle of permission or connectedness. Each of these methods naturally assumes a common ground since methodologies always reflect a moral view. The question that emerges, however, is how we identify the common ground. A second task is to see what this commonality tells us about methodology: what is lacking is a philosophical understanding of method.

In view of the limitations related to most methods, it might be worthwhile to try an entirely different avenue for finding common moral ground: *proceduralism*. Some, most notably H. T. Engelhardt, Jr., have argued for a *procedural* understanding of bioethics as a way to resolve the contentious issues of bioethics in secular societies. In the midst of the particularities of moral communities, Engelhardt argues that a secular society can identify common procedures that allow people who are moral strangers to collaborate peaceably in health care. These procedures are built on consent and permission. Proceduralism provides a minimal grammar for moral discourse, but it also reflects common moral commitments that are often overlooked. Engelhardt has treated his proceduralism, as have many of his critics, as if it were content-thin. However, Engelhardt and his critics may have under-estimated his discovery. He has not explored the idea that a procedural bioethics must also be justified on common moral assumptions (e.g., the dignity of persons, the importance of freedom, or the value of peace in a peaceable society).

A crucial criticism of contemporary liberal democracy is that it focuses on procedures without any attention to the moral consensus necessary for its procedures to work. The danger in this situation is that the ethical nature of political freedom will be lost and the pro-cedures of liberal democracy will lack any moral justification.[1] How-ever, rather than search for an extrinsic moral consensus, one can look to the procedures themselves as initial points of agreement. One might think of proceduralism as first-order moral discourse or prac-tice. Bioethics can be second-order discourse that analyzes the first-order practices.

This turn toward procedure offers an alternative way to view bio-ethics. It avoids the fragmentation of communitarian particularism and the seeming emptiness of liberal proceduralism. A communitarian view often treats complex and diverse secular societies as a commu-nity and fails to take seriously the differences between moral commu-nities (see chapter 5) or falls into some form of communal relativism. The liberal procedural view sees all particularities as difference with nothing binding the communities together. However, this view can-not account for the commonalities and overlapping points of consen-sus found in secular societies.

The examination of proceduralism in bioethics assumes that secu-lar society is not a community or a mere aggregate of diverse views. Rather, secular society is a collection of communities with different interpretations of the moral life and the common good. While proce-dural resolutions have been important to resolving moral dilemmas in health care, little attention has been given to *procedures as moral practices* that embody certain moral commitments. The use of proce-dural ethics, which has been so important to the contemporary prac-tice of medicine and health care, challenges the field of bioethics to

examine the moral assumptions underlying common procedural resolutions. The practice of procedural resolution also points the field in
another direction. That is, bioethics needs to pay more attention to
organizational and institutional ethics. The questions of bioethics in
the clinic, in the research laboratory, or in the public forum are set in
a context of organizations and institutions. These contexts shape the
choices that can be made. This chapter will argue that bioethics must
develop a methodology that analyzes institutional arrangements and
practices.

—— MORAL ACQUAINTANCES AND PROCEDURAL BIOETHICS ——

In the midst of different views about method and content in bioethics
are some who argue, Engelhardt most notably, for a procedural methodology. That is, substance (or content) should be ignored and the focus
of ethical reflection should be on methods (procedures) for resolving
moral controversies. The procedural argument is that moral pluralism
creates such diversity of views that only the moral authority of procedures (most notably procedures of permission and agreement) can
span the differences. Procedures such as informed consent are held to
illustrate the creation of moral authority in the face of moral diversity absent the authority of a shared moral vision. The argument by
Engelhardt is that a procedural methodology can span different moral
communities because procedures are content-thin. A result of this argument is that there are two species of bioethics: the bioethics of
different communities; and minimalist, procedural bioethics for secular societies.

   This division is, however, too simple. Women and men, from different communities are able to reach agreements at different times. They
can engage in discussion and give reasons to one another. Further,
most men and women do not live in a single moral community but
in many. Even the most minimalist procedural view of bioethics must
assume a world of moral acquaintances (see chapter 5) if the procedures themselves are to be morally justified. There are clear limits in
appeals to such agreements. In chapter 6, we argued that such agreements must be approached with caution as they are often more fragile
than they appear. The advantage to the language of moral acquaintanceship over the language of common morality or overlapping consensus is that it gives a better account of the fragile nature of the

common agreement and the greater strength of the agreements in other areas.

The language of common morality and overlapping consensus indicates a terrain of moral agreement. But that terrain is not clear cut. Much of the agreement found in bioethics is fragmentary and ambiguous. For example, for different reasons, strong agreement surrounds the procedure of informed consent. But that agreement can break down, as in the controversies over physician-assisted suicide. It breaks down because the agreement about self-determination and other values underlying informed consent are not so clear cut. The metaphors of a "common" morality, "connectedness," or "overlapping consensus" are often misleading. At the same time a two-tiered division of moral friends and strangers can also miss the point that moral content is needed if procedures are to be morally justified. Without such justification, procedures simply become the assertion of power. Those who argue for a "pure proceduralism" or "minimal proceduralism" fail to recognize that the choice of a method—even a thin procedural method—involves some commitment to moral content. Method and content cannot be discussed in isolation from one another. The separation of method and content reflects a Cartesian heritage that has sought to identify pure method without content.² A conceptual distinction between procedures and content can be drawn, but the distinction must be recognized for what it is. The two define one another; they do not exist apart from each other.

The most basic source of moral authority to which one can appeal is the authority given by moral agents through agreement. This justification is minimalist and procedural. It is an appeal that can be understood and shared without sharing a common understanding of the good or a content-full understanding of the right. Moral strangers can understand whether or not permission has been given.

In the appeal to procedural bioethics, contemporary secular practices provide instructive examples of authority by mutual consent and agreement. Limited democracies, with limited authority and exclaves of privacy, have their moral authority grounded in the consent of the governed, and the free market provides innumerable examples of authority, grounded in consent, which can be understood by moral strangers. The practice of free and informed consent in health care is another example of moral strangers collaborating with moral authority grounded in consent. Indeed, many of the achievements of bioethics (e.g., informed consent, advance directives, prior notification, futility policy) are examples of this procedural morality. These procedures are not empty, or even thin, in terms of moral content. They rely on assumptions about the value of human beings, human integrity, and honesty.

Some, mistakenly, will label the minimalist appeal as simply one more liberal appeal to individualism. That is, critics often assume that appeals to proceduralism and consent assume liberty as a primary value. But the turn toward proceduralism need not be an appeal to individualism, freedom, or autonomy. Such appeals involve ranking some values over others, but results thus reached are only one among numerous particular accounts. What is wanted is a transcendental argument that sets out a minimal grammar for moral discourse when men and women do not share the same substantive views of moral justification or content-full views of the moral life. Robust, thick views of the moral life are developed within moral communities that share common moral frameworks (moral friends). In a political society that understands itself as secular, there are limits on the moral vision, health care policy, and content of bioethical conclusions that can justifiably be imposed by the state. Such limits stem from the limits of our common moral knowledge. We can only justify what is shared. What we can always share is agreement. Common projects, like health care, are possible insofar as they are freely undertaken.

If procedures are not as empty as supporters and critics of procedural morality think, the exploration of proceduralism can further the discussion of moral acquaintanceship as such a category for secular bioethics. In proceduralism one discovers some very helpful points for bioethics insofar as these procedures indicate a terrain of overlapping values that can be understood and shared by moral acquaintances.

Often secular societies may share dominant, homogenous cultures and traditions that foster overlapping points of view in bioethical discussions. Many European bioethicists appeal, for example, to the principle of solidarity to frame bioethical questions. They are able to make this appeal because they conceive of secular society as a homogeneous culture rather than as many cultures in a society. The principle comes from a common heritage of religious, cultural, and political traditions. In more pluralistic societies, such as the United States and Canada, where there is greater cultural and moral diversity, there is less to share in common for bioethics.

BIOETHICS, MORAL ACQUAINTANCES, AND THE
COMMON MORALITY OF PROCEDURALISM

A purely procedural society—and its purely procedural bioethics—is confronted with a conceptual problem. There are no "pure," content-less procedures. Following Wittgenstein's observation that rules are part of the ways of life, one can argue that procedures reflect political, social, and moral assumptions. Procedures, no matter how sparse they may be, reflect a set of shared moral commitments. Without a shared moral view—no matter how thin—procedures cannot be morally justified and are simply imposed by the power of a particular point of view. Without a shared view, procedures can only be sustained by the use of power.[3] Insofar as we lack a neutral procedural view, we must turn toward a procedural view that explicitly acknowledges its underlying assumptions. By acknowledging the moral assumptions within proceduralism, we can identify a moral common ground that allows us to critique procedures and methods in bioethics.

What are the practices and the corresponding moral assumptions and commitments that must be shared in a secular, pluralistic society? Among the key procedures of bioethics are procedures of consent (see chapter 1). When there are different views of the good life or what constitutes appropriate health care, consent becomes crucial for the moral justification of cooperative action. One of the conditions for achieving the good of self-government is a commitment to liberty. This basic commitment to liberty leads clearly to a commitment to a principle of consent. The consent of the governed is the moral commitment that defines the source of authority for the state; it is also essential for moral authority in bioethics. Consent creates agreement and makes cooperative endeavors possible. However, we can only have practices of free and informed consent if we share assumptions about the freedom of individuals, honesty, trust, and other assumptions about the human person often expressed in terms of liberty and human dignity. Consent also reflects a moral commitment to respect individual conscience and moral integrity. It is also reflected in procedures that protect health care institutions and the conscience of individual health care workers.

A second practice of bioethical proceduralism is the rule of law providing a fundamental framework for secular society and bioethics. One finds manifestations of this rule in the practice of free and informed consent in clinical treatment and research. The development of futility policies provides another example of the rule of law insofar as institutions have the responsibility to determine how they will or will not use resources. Another practice grounded in the rule of law is that of prior notification in which an institution must tell a patient, or the patient's surrogates, in advance, of the content and method for its policies.[4] The more an institution can articulate its mission and contextualize its procedures, the better the procedures will work. Such statements and guidelines set parameters of expectations, protect individuals and institutions, and foster respect and toleration. Nevertheless, deep disputes will still arise about what the rule of law can allow or prohibit, as is evident, for example, in the debates on abortion or physician-assisted suicide.

While some will complain that the law's disclosure requirements overly complicate and bureaucratize medicine, such requirements can also be understood as ways to protect and maintain the moral integrity of patients, physicians, health care workers, and institutions. Such procedures are ways to insure that all similar cases are treated fairly (i.e., in the same way). Indeed, one of the underlying values that support such practices is an overlapping commitment to fairness and impartiality. One can argue that procedures have taken on an increasingly important role in bioethics and ethical questions in general, insofar as the context of ethical questions has changed. That is, in contemporary society, human life is increasingly contextualized and embedded in institutions. Health care provides a myriad of examples of institutional life: hospitals, managed care organizations, physician practices. Ethical questions are not exhausted by asking: What should I do? Ethical questions and issues need to be analyzed in the context and fabric of institutional settings.

A third practice implicit in morally pluralistic, secular societies recognizes the limits of the moral authority of the state and the distinction between the state, society, and community. One might distinguish the state from society by assuming that the state is concerned with the order of politics while society is concerned with the order of culture. The state has a more restricted and defined sphere.[5] The moral pluralism of postmodernity limits the moral authority of the state.

From the betrothal of church and state by Constantine through contemporary legal decisions,[6] the state has been understood by many as the protector of public moral culture.[7] In his commentary on

English law, Sir William Blackstone speaks of the natural law that can be discovered by reason. The laws of nature are "the eternal, immutable laws of good and evil, to which the creator himself in all his dispensations conforms, and which he had enabled human reason to discover, so far as they are necessary for the conduct of human actions."[8] Blackstone argues that the law of nature, the moral law, is one of the foundations of the civil law. The civil law builds on the moral duties one has to God, neighbor, and the self. It has a remedial function when people violate natural law or fail in their moral duties.[9] This view of the state presumes that society knows what is good for each person and that it has the moral authority to enforce it.[10]

These assumptions are challenged by the moral pluralism in contemporary secular societies and argued in bioethical controversies such as abortion or physician-assisted suicide. These subjects are controversies twice over. First, they are controversial as to their moral character. Is it morally appropriate to participate in an abortion or to assist someone in taking his life? Second, such issues are also controversial as to their understanding of the proper moral function of the state. Is it the state's moral duty to prohibit abortions or assisted suicide? May the state enforce a particular resolution to bioethical controversies? Issues in bioethics challenge the degree to which the state can be the keeper of public morals. The controversies of bioethics reveal the extent of moral pluralism and the limits of state power that can be plausibly identified as good for public morality.

Simply because the state is limited in its authority to enforce morality does not mean that there is no "common" morality that shapes the culture of a particular secular society. The public moral culture is a domain of discourse for moral acquaintances. It is instructive that many who posit a common morality (e.g., Beauchamp and Childress; Gert, Culver, and Clouser) understand it as open to many interpretations and development. Thus, appeals to common morality, as part of a bioethical method, can be diverse. For example, the secular state tolerates an array of peaceable behaviors that many men and women find offensive. Such tension is obvious in debated issues such as abortion or the use of RU-486.

In another debated issue, I find no general, secular moral argument that can justify the state's prohibition of assisted suicide. Arguments against suicide are unavoidably based in particular views of the good life and the good death. The state cannot determine how to balance the risk of dying too early against the risk of suffering too long. Authority to resolve such controversies will be available, in secular terms, only from the person involved. Assumptions about the sacredness of life or the meaning of suffering that are not shared cannot provide general secular moral arguments against suicide. Many of the most interesting arguments opposed to physician-assisted suicide turn on questions of how physicians, as a profession in society, ought to function.[11]

While the moral authority of the state is limited, the state is nevertheless crucial to a secular society. It provides a framework in which moral strangers and acquaintances, those with differing and diverse moral points of view, can meet and cooperate peacefully. The state remains central to protecting the rights and exchanges of moral agents. It can punish those who unjustly take from others by force or deception. It can enforce agreements that have been freely recorded. The state is the appropriate agent for allocating commonly held public resources and publicly created goods; however, the state is appropriately neutral toward all particular and peaceable understandings of the good and of the content of right conduct. Because we do not have a content-full understanding of good public morals, the society served by the state is not one community but many communities with diverse bioethics.

The distinction of state, society, and community leads to a fourth element of a procedural bioethics: toleration. The moral diversity of secular societies means that there are substantive moral positions that some men and women do not approve. To practice toleration is to recognize the limits of moral views and knowledge insofar as everyone cannot reach the same conclusions. It is, in short, to recognize that there are differences in moral experience. If morality is part of a way of life, then different experiences can lead to different ethical reflections. Toleration is the practice that recognizes the tentative and provisional nature of our moral judgments and the limits of our moral knowledge. Toleration is, finally, a practice that supports moral integrity. It allows men and women with distinct views to maintain their moral values and integrity in their moral lives.

Current discussions regarding assisted suicide and euthanasia point to the importance of toleration and the difficulties faced by those who wish to avoid engagement in practices that they recognize as morally wrong. In developing a "protocol" for physician-assisted suicide Timothy Quill and his co-authors acknowledge that this action may be offensive to physicians with particular content-full moral views. The protocol recommends that physicians who take offense should refer patients to someone who will not. The authors recommend:

> No physician should be forced to assist a patient in suicide if it violates the physician's fundamental values, although the patient's primary physician should think seriously before turning down such a request. Should a transfer of care be necessary, the personal physician should help the patient find another more receptive physician.[12]

The protocol misses the point that a physician may find engaging in the referral itself to be morally wrong and a formal cooperation with evil.

In the case of physician-assisted suicide there are two important lessons about toleration. First, the conflict of moral visions and the demands of moral integrity provide a reminder about the proper use of state authority to enforce a particular moral point of view. For example, it is one thing for a state not to restrict physician-assisted suicide, and quite another for a state to require participation in finding someone who will aid in suicide. Further, it would be coercive interference to require that the state fund assisted suicide or mandate such services as part of everyone's basic health insurance or in the training of physicians.

Second, physician-assisted suicide helps illuminate the confusion between toleration and acceptance.[13] Many people misconstrue toleration as entailing the approval of or acceptance of immoral actions and lifestyles. Political and secular moral offense is often taken, if a lifestyle is morally condemned,[14] on the grounds that toleration of differences requires acceptance. Such offense stems from a failure to understand that toleration does not require approval but in fact involves disapproval. One tolerates what one dislikes, disapproves, condemns, or hates, though one eschews coercive force in one's disapproval. Misunderstanding toleration as approval can lead to a requirement of toleration as a central virtue. It then becomes a central element of a coercive ideology that all lifestyles must be treated as matters of choice. This view is a moral view yet its enforcement lacks general secular moral justification. A secular society has secular moral grounds for reclaiming the original meaning of toleration, which then becomes the linchpin of secular public policy in general and of health care policy in particular. Secular policy must recognize that individuals and communities are morally free to condemn peaceably that which violates their moral view. Christians may tolerate the sin of abortion in secular society but still condemn the behavior and pray for society's repentance.

Given the limits of secular moral authority, the state's commitments must be recast in favor of greater liberty. Space must be made for a peaceable moral diversity that will not require morally compromising collaboration in evil or seek to enforce a political correctness that would compromise the integrity of peaceable moral communities. Taking moral diversity seriously requires letting it exist and letting moral judgments be made. To take the diversity of moralities seriously is to confront peaceable conflict among differing moralities. If such conflicts exist in a world in which moral orthodoxies are coercively compromised and political correctness enforced, confrontation is impossible and some version of the Amish option becomes the only attractive alternative: a social order in which communities may develop their own schools and health care systems that reflect only their own values and moral understandings. One can then imagine a society that develops something like a voucher system to enable men and women to participate in publicly created goods, like health care or education, without compromising their moral views. Members of the pluralistic secular society would use the "vouchers" to procure their share in the public good, for example, health care, but simultaneously using their share according to their own moral commitments. Different communities could, in good conscience, contribute to the creation of the public good and use it according to their own moral commitments. Men and women could use such vouchers in ways they judge to be morally appropriate. For example, two people facing death may deploy health care vouchers in different ways. One may use vouchers to sustain life as long as possible; the other may use them to visit Dr. Kevorkian's clinic. In this way, both would have access to a public good in a way that accorded with their moral vision. In some ways, such a system would represent a contemporary version of the Ottoman Empire's *millet* system.[15]

If a society views health care as something necessary for human existence (i.e., as a right or as something a morally good society ought to provide), then it will have to address, structurally, how to support health care and tolerate differences. The voucher model previously discussed is one way to address such problems. In terms of health care, this model would allow citizens to make health care choices congruent with their own moral views. It would also allow institutions and health care professionals to practice health care delivery according to their moral views.

Reflection on proceduralism in bioethics provides important points for further exploration. The procedures of free and informed consent and institutional identity, to name but two, rely on overlapping views of freedom, honesty, and respect for persons. The rule of law assumes a commitment to impartiality and fairness. The practices of toleration evidence underlying views about the value of moral integrity and the limits of moral knowledge.

In a procedural method, we can identify points of overlapping agreement for moral acquaintances, and we can find ways to frame and talk about disagreements. A turn to procedural structures creates the possibility for civil conversation in bioethics. It carries an underlying commitment to inclusion in that no one is, a priori, excluded from the civil conversation. Such conversation, of course, poses both risk and possibility for moral communities. Conversation with others may lead to the growth or development of a tradition of moral thought that influences the views of others. The risk is that a community may lose its identity.

Procedural ethics provides a way in which the terrain of moral acquaintances can be identified and helps us understand the roles of different methods. The strength of the acquaintances and the context of the moral issue will influence the choice of methods. For example, a moral controversy within the clinic of a religious institution may allow for the use of casuistry. However, if the moral acquaintanceship is less strong, the conversation may best begin using some form of principlism. Methodological reflection in bioethics must identify the common ground of moral acquaintances and the context of the moral issue at hand. If procedural ethics is a key for developing the terrain of moral acquaintances in bioethics, we cannot limit this exploration to procedures that involve patients. The field will also have to turn its attention to organizational and institutional ethics as these are the contexts of our procedures.

——— ORGANIZATIONAL ETHICS ———

Until very recently, the focus of bioethics has either been on individual situations and cases or on the level of national policy and law. However, if one attends to proceduralism as a way to identify and understand the common moral ground of moral acquaintances, one becomes more and more aware of the place of intermediate organizations and institutions in health care.[16] Such institutions fill the gap between individuals and the nation. They include hospitals, health maintenance organizations (HMOs), insurers, delivery networks, research institutions, and the like. These institutions and organizations frequently form the context for any moral question. The institutional context shapes the moral question. The role of such organizational contexts are very evident in managed care as these organizations (i.e., MCOs) mediate between the individual patient and the larger society.

The developing nature of institutional and social structures that support health care prevents the field of bioethics from asking ethical questions as if they only existed at the level of individual choice. Many moral issues that confront individuals are well beyond the reach of individuals: many are systemic or social issues. Thus bioethics has moved beyond the bedside, the clinic, and the hospital.

It may help my argument to begin by putting the present American health care situation into a broader historical context. Until recently, medicine has been both a product of, and resistant to, modernity. Now, the shifts to models of managed care and corporate medicine have moved important parts of health care from the guild model of the late Renaissance to the corporate model of modernity. This context and change creates a new composite of the place for ethical questions. It is in this context, too, that new questions arise, such as how one can and should think about institutional identity and institutional conscience.

MEDICINE AND MODERNITY

There are many ways to talk about modernity, and recent debates about "postmodernity" illustrate how difficult it can be to define modernism and the modern age.[17] Nonetheless it seems fair to characterize modernity, in part, as a shift from the particular to the universal. In the modern age, science has been the dominant model of knowledge. It has replaced theology, the model of the Middle Ages, as the central method of human knowing. In the modern age even theology aspires to be "scientific." In modernity, scientific knowledge is characterized as universal and objective. At the same time, modernity has seen the evolution of the centralized state and the rise of bureaucratic structures to address basic human needs (welfare systems).

Changes in the economic order (e.g., industrialization/technology, capital markets) have led to more bureaucratic structures in social life. Systems of education and health care arose not only in response to the explosion of knowledge but also to replace what families used to do but could no longer do in the urbanized economic order.

Until recently, medicine as practiced in the United States has had a peculiar relationship to modernity. On the one hand medicine has benefited from the rise of the scientific model. Scientific knowledge redefined the purpose of medicine as "cure."[18] Hospitals reflect the rise of bureaucratic structures to deliver health care, but physicians have retained a guild/artisan model of operation. The emphasis has been on solo practitioners. Even hospitals were largely stand-alone operations.

The development of managed care in the United States continues the modernization of American medicine. While the surge in managed care was driven by the costs of fee-for-service medicine, its model of care is tied to modern, bureaucratic, corporate life. Managed care represents the completion of a shift of medicine from the guild model to the model of corporate/bureaucratic health care. It represents the entry of heath care into the modern age; the change in economic structure completes the transition.

The change to the model of managed care focuses the practice of medicine on the care of populations rather than the individual patient. To view medicine as addressing the needs of particular patients is an incomplete reading of the modern context. Managed care forces physicians to see themselves as part of larger teams. Furthermore, managed care brings the fruits of scientific study and knowledge to medical practice (e.g., evidence-based medicine, outcome studies, patient assessment).

A NEW CONTEXT FOR BIOETHICS

The traditional context for questions of ethics has been a focus on particular individuals. The same focus characterizes bioethics: What should I do? How should the patient be treated? What is my role as a health care professional in this situation? However, the shift to managed care adds a new level of ethical discourse and puts the individual's questions in a different context. The physician is no longer the solo practitioner. The hospital is no longer a stand-alone facility.

Institutions and professionals are set in a maze of relationships and obligations.

How might we begin to frame this new or additional dimension for ethics? I think we can draw on two resources. First, Catholic moral theology has much to teach us about institutional identity and cooperation. Second, the tradition of Roman Catholic social thought and ethics has raised important questions about the structures of society.

The moral identity of an institution, that is, its mission, is the first element or necessary condition for talking about institutional conscience. How does the institution understand itself? How does it present itself?[19] A mission is not comprised exclusively of moral values and commitments. But its moral commitments should be identified in its mission.

Within the broad institutional parameters of its mission, one can begin to think about an institution's moral integrity. That is, does it live up to its moral commitments? A lived commitment is more than the achievement of a simple unimaginative consistency. As circumstances, contexts, and problems change and develop, so too must institutional integrity involve creativity and fidelity. The institution that develops an understanding of its mission has the criteria that can be used to develop the tools to evaluate mission effectiveness and shape and implement institutional conscience.

Consider, for example, the Ethical and Religious Directives (ERD) for Roman Catholic health care institutions in the United States.[20] These directives set out a broad framework in which Roman Catholic institutions can articulate their mission. The ERD are best known, perhaps, for the prohibitions they contain (e.g., against abortion, euthanasia, sterilization). Equally important, however, is the positive vision they articulate for Catholic health care institutions. One can argue that Roman Catholic health care should be committed to the care of the poor—not only the economically poor but all those who are weak and vulnerable. Such institutions will also be concerned with the stewardship of resources as well.

TOOLS FOR LIVING OUT THE IDENTITY

Assuming that an institution or system articulates a mission, how does that mission become implemented? How can a vision become real? It is essential to the development of institutional conscience that, whatever form its mission takes, everyone in the institution must have ownership in that mission.

Another critical tool for the integrity and conscience of an institution is its *budget*. A budget can be understood as a planning document or as an articulation of the institution's mission. For example, if an institution says in its mission statement that it is committed to care of the poor, but makes no line-item commitment to this care, then the institution is either in deep self-deception or it is lying.

In addition to a budget, an institution's strategic plan and planning process are important to mission, identity, and conscience. Just as individuals plan according to life goals and objectives, so too health care institutions and systems need to deploy a planning process. A measure of the conscience of an institution is the degree to which the institution's moral commitments are part of strategic planning. They ought to shape the institution's long-term goals and the means that are used to achieve them.

The process of planning involves a process of self-study and evaluation. One can well imagine having a hospital ethics committee, or an institutional analogue of an ethics committee, review the practices and policies of the institution. The focus would be on patients and particular issues of clinical care but also on the whole moral culture of the institution (e.g., its treatment of human resources, financial planning, and advertisement).

In this whole process the role of the trustees or directors is crucial. They have a special responsibility to call and lead the institution in fidelity to its mission and a special role in the articulation of institutional mission and identity. While others may be more concerned with the day-to-day details of patient care and institutional management, the duty of the trustees is to ask for accountability for broader questions of institutional identity and life.

The development of an institutional moral identity and conscience depends then on an ongoing process of education to shape the culture of the institution (recall Captain Cook and the Hawaiians). Education is important for ongoing renewal and adaptation, but it is also important for consent. Members of institutions—patients, workers, and professionals—need to know and consent to the mission. They need to be aware of the culture of the institution they are joining. Of course, there is rarely a "perfect fit" of individuals and institutions, so another crucial tool will be provisions for the protection of individual conscience.

SOCIAL THOUGHT

If bioethics attends to questions of organizational and institutional ethics it can also attend to questions of social justice.[21] My use of this term is influenced by Roman Catholic moral thought which distinguishes between distributive and social justice. While distributive justice is concerned with how common goods (e.g., wealth or health care) are distributed, social justice is concerned with how a society is organized. For example, a social justice perspective on the delivery of health care services in the United States would be puzzled by the organization of medicine. Here health care is developed from both public and private sources and investment. Public investment supports medical research, education, and the building of medical facilities, yet there is no assured access for all citizens to services that are developed by public investment. The complex issues of distribution and the organizational lens of social justice raise important questions of organizational ethics that must be addressed.

———— Summary ————

Bioethical issues have two different moral dimensions. The first moral dimension is the question or issue itself. Should the patient be intubated? Should physicians assist patient suicides? How should we allocate health care resources? What types of research should we fund and how should research be conducted? How should we deal with developing genetic technologies? These are typically the issues of bioethics. They have not been, however, the primary focus of this book. This book has investigated the second dimension of these questions, namely, the fundamental philosophical questions raised by the issues. Such an exploration allows our conversation to move beyond the particular questions of bioethics to the issue of methodology. The exploration of method leads, in turn, to the broader issue: a reflection on practicing ethics in a secular, morally pluralistic society.

The history of bioethics can be read as a growing awareness of the implications of living in a secular, morally pluralistic society. Having grown beyond physician ethics and the reflections of different, usually isolated, theological and moral communities,[22] bioethics has become a philosophical discipline with an emphasis on legality, consensus, and public policy. This development, in a sense, is only natural as law and philosophy are essential disciplines for all discussion in secular society. But as philosophical methods seek common ground, they also downplay multiculturalism and moral pluralism. The emergence of many voices in bioethics challenges the assumption that we do really have a common morality or even a common moral rationality. If listened to attentively, these voices raise the question of whether there is a unified practice of bioethics. Our conclusion can be baldly stated: rather than begin with an assumption of a common method into which diverse voices can be placed, we must begin by listening to the diversity that is already there. Only so can we discover common ground.

The exploration of method leads us to conclude that just as there is no singular voice or language for secular bioethics, so there is no singular method in bioethics. We have, rather, many methods that depend, in part, on the setting (e.g., the clinic, public policy) and the principle discipline one favors (e.g., law, theology, or philosophy), and finally, methodological choices depend, in large part, on the assumptions one makes about the moral world. As a field of inquiry and reflection that seeks to address ethical questions in the context of a secular, morally pluralistic society, each choice of method in bioethics reflects a set of moral commitments and assumptions. Method is a way to articulate a view of the moral world or one's moral experience.

If one articulates a methodology, a philosophy of method for secular bioethics, one would have to take an eclectic stance. No one method can secure the field because different settings and contexts—the clinic, the laboratory, or the public forum—pose different issues and dilemmas, and different views of the moral world bring different methods to bear. It is not surprising that the moral world is furnished in many different ways. Each method is a partial expression of the moral world.

What, then, ought the field of bioethics do? One can argue that it should fill at least four roles. First, the bioethicist is a member of a particular community, and as an authority within that community, he or she can work to help the community understand new challenges presented by contemporary health care or find ways of speaking and acting in the public forum. The bioethicist functions as an authority on the tradition of the community. Here the bioethicist is most like a priest, rabbi, or theologian in a community.

A second role for a bioethicist is that of a translator. That is, while different communities may speak different moral languages, people can learn to speak many languages. So too, there is no reason, in principle, that bioethicists could not speak different moral languages and understand different views and methods. The bioethicist can work to enable communities to understand their differences and similarities with others. In the debate about physician-assisted suicide, for example, the bioethicists can help different groups understand how others view the importance, or lack thereof, of intention as part of the moral analysis.

The third role is that of a geographer of authority and ideas. The bioethicist can help to trace the lines of authority and permission in the practice of medicine and health policy. The secular context is one of consent, permission, and understanding. The bioethicist can also help the community understand the views of other communities on different topics. People often live in several moral communities. They often live in the midst of fragmentation and incoherence. One role for secular bioethics is to help sort out the different, and sometimes incompatible views, that people may hold.

A fourth role for the bioethicist is to work with others to analyze institutions, organizations, and organizational ethics. Are institutions faithful to the mission they profess? Are they honest in the agreements that they make? How do they deal with the men and women who seek health care from them? How do they deal with the women and men they employ? In this role, the bioethicist also functions as a social critic.

The realm of procedural bioethics provides a basis for bioethics as a field to explore common morality in a secular society. However, as work is done on the procedural level, bioethics will be involved unavoidably in the institutional and organizational level of ethics. Though there is a growing awareness of the institutional and organizational dimensions of bioethics, the realm of organizational ethics is often overlooked. All too often ethics seems to work between two extremes: the individual and the whole. What should be done in this particular case? Or, what should the law or public policy be? However, a characteristic of modern life, especially modern medicine, is that human beings are embedded within intermediate structures. Bioethics is not simply about the individual patient and that patient's physician. Nor is bioethics exhausted by national health policy or law. Rather, there is a collection of structures and organizations—hospitals, insurers, providers, families, churches, communities—that form the contexts for discussions in bioethics. Choices are often not just those of the patient or physician. Rather, choices are embedded in a moral context that includes institutional interests and integrity. An important work of bioethics is to identify and analyze this web of institutional and individual relationships.

—— Conclusions ——

Health care touches all dimensions of human life—sexuality, conception, birth, illness, money, and death. It also touches the life of society to the degree that it deals with questions concerning the allocation of common resources, the use of public goods, and the type of society people hope to live in. As such it provides a context in which men and women often meet in moral diversity.

Health care provides examples of moral pluralism and examples of how men and women can act together with moral authority, as moral acquaintances, in the face of moral diversity. Moral communities are particular in history and vision. To understand their substance and character is to see them in tandem with other possible moral communities and as the basis of the general secular morality that can best be sketched in terms of permission or agreement. Particular communities are not necessarily set in isolation from one another. They can engage in discussions and arguments insofar as they share sufficiently in commonly held substantive moral commitments and justification.

Free and informed consent, advance directives, and the prior notification of a physician or an institution's moral commitments are procedures through which moral strangers and acquaintances can work together. Moral acquaintances can work together to address specific issues, and as we have seen repeatedly, agreement on procedures reflects some degree of agreement on the underlying values necessary to support the procedures. Exploration of procedural bioethics is a good direction to take in the effort to better understand and evaluate the common ground of moral acquaintanceship and the procedures themselves. The more we understand the common ground of moral acquaintanceship, the more order we can bring to the anarchy in bio-ethics, as Wickler has described the methodology.[23] It will not be a well-ordered hierarchy or even a kind of lexical ordering, but the common ground shared by moral acquaintances will let secular bio-ethics better critique its choices of method and procedures.

Under examination, it is possible that secular bioethics can help us navigate moral controversies insofar as we can delineate the lines of moral authority. Secular bioethics helps map the geography of moral commitments, language, and ideas. Substantive bioethical reflection, about how men and women ought to behave, is clearly the work of bioethicists within particular traditions. Within moral communities bioethicists serve to maintain the moral integrity of the community. Bioethics can also be a work in the domain of moral acquaintances where there are overlapping moral commitments around certain issues and values. At the secular level, such a bioethics can help identify the common ground between moral communities. This common ground addresses issues in health care but also opens a window for other moral issues in society.

Secular bioethics offers a framework in which moral communities can flourish and establishes a bulwark against an unqualified relativism. Above all, however, it also provides a way for moral strangers to understand moral authority without recourse to force. While people may understand one another as moral acquaintances, it is the appeal to the moral authority of permission that provides the framework for the resolution of moral controversies among those who are not moral friends. In the search for methodology we have found that method and content are part of a package. Any method, including the method of proceduralism, reflects basic shared moral commitments. Procedural bioethics must make certain assumptions if the procedures are to work. These are the overlapping agreements of moral acquaintances. There must be a commitment to honesty in informed consent or a common value of peace if permission is sought.

In one sense, bioethics is a confrontation with the limits of human life. Clearly, in spite of our increasing technological capabilities, the practice of medicine daily confronts the limits of human existence. Human beings die; they get sick; they suffer. Often, there are insufficient resources to help. Bioethics is a confrontation with the limits of human knowledge about what is right and wrong in making decisions for life and at the end of life. There is often a tendency to overlook or downplay the limits of our knowledge. Yet it is only in knowing the limits that we can come to know the real possibilities of the field.

In our ending is our beginning.[24] Secular bioethics emerged from the work of moral theologians and other moralists working in particular religious traditions.[25] Many of those thinkers, however, did not see themselves as working with a particular morality but with a particular expression of morality. In the last thirty years, secular bioethics has followed the history of philosophical ethics. It has sought to move beyond the boundaries of particular communities to establish a content-full bioethics on the secular level. In so doing, however, bioethics has often assumed that a common content-full secular moral community exists. To some degree that assumption has failed. The desire for a universal method does not address the particularities of moral communities. In a return to moral communities, secular bioethics now has the task of providing moral geography, tracing the lines of permission, and doing institutional analysis. Secular bioethics can help to chart the boundaries of reason, content-full moral language, and the authority of the state and health care institutions. That is, in its turn to moral communities and the notion of moral acquaintances, secular bioethics finds itself in a position, through a procedural analysis, to find the common morality that can ground and justify its procedures.

NOTES

---◆---

——— Introduction. Methodology and Bioethics ———

1. For a wonderful history of the field see Albert R. Jonsen, *The Birth of Bioethics* (New York: Oxford University Press, 1998).

2. Daniel Wickler, "Federal Bioethics: Methodology, Expertise and Evaluation," *Politics and the Life Sciences,* February 1994.

3. Recently the journal *Theoretical Medicine* has begun a regular section on methodology. See E. Erde, "Method and Methodology in Medical Ethics: Inaugurating Another New Section," *Theoretical Medicine* 16: 235–38 (1995). One also finds serious questions being raised by those outside the field as to what it actually contributes to the questions that are raised. See Udo Schüklenk, "Time to Rethink Bioethics Method," in Letters in *The Bulletin of Medical Ethics,* May 1995, and Dr. Yarman Ors, "More on Bioethics Method," in Letters in *The Bulletin of Medical Ethics,* September 1995.

4. Throughout this book I will argue a position that is not committed to moral relativism. I believe there are objective standards of morality. The question I am trying to confront is how to overcome epistemological difficulties in knowing these standards.

——— 1. Bioethics and Moral Acquaintances ———

1. In a recent book Erich Loewy has introduced the category of "moral acquaintances." I would argue that he sees the common ground of moral acquaintanceship in a much stronger way than I do. See chapter 5. Also see Erich H. Loewy, *Moral Strangers, Moral Acquaintance, and Moral Friends: Connectedness and Its Conditions* (Albany, N.Y.: SUNY Press, 1997).

2. For an excellent history of the emergence of bioethics see Albert R. Jonsen, *The Birth of Bioethics.*

3. Ludwig Edelstein, "The Hippocratic Oath: Text, Translation and Interpretation," in *Ancient Medicine* (Baltimore: Johns Hopkins University
Press, 1967), p. 55.

4. See Edmund D. Pellegrino and David C. Thomasma, *A Philosophical
Basis of Medical Practice* (New York: Oxford University Press, 1981); *For the
Patient's Good: The Restoration of Beneficence in Health Care* (New York: Oxford
University Press, 1988).

5. See, for example, the Park Ridge Center series on different religious
traditions and health care.

6. Joseph Fletcher, *Moral and Medicine* (Boston: Beacon Press, 1960);
Edwin F. Healy, *Medical Ethics* (Chicago: Loyola University Press, 1956);
Immanuel Jakobovits, *Jewish Medical Ethics* (New York: Block, 1958); Paul
Ramsey, *Fabricated Man* (New Haven: Yale University Press, 1970); Gerald
Kelly, *Medico-Moral Problems* (St. Louis: Catholic Hospital Association, 1958);
Harmon L. Smith, *Ethics and the New Medicine* (Nashville: Abingdon Press,
1970); James M. Gustafson, *The Contributions of Theology of Medical Ethics*
(Milwaukee: Marquette University Press, 1975); Richard A. McCormick,
Health and Medicine in the Catholic Tradition (New York: Crossroad Press, 1984).

7. Edmund Pellegrino, "The Metamorphosis of Medical Ethics: A 30-
year Retrospective," *Journal of the American Medical Association* 269: 1158–62.

8. See Kathryn M. Olesko, "Technology, Efficiency, and Gender in
Evangelium Vitae," in *Choosing Life: A Dialogue on* Evangelium Vitae, ed.
K. Wildes, S.J., and A. Mitchell (Washington, D.C.: Georgetown University
Press, 1997), pp. 107–22.

9. T. S. Eliot, *Sweeney Agonistes,* "Fragment of an Agon," in *The Complete Poems and Plays* (New York: Harcourt, 1958), p. 80.

10. See Gilbert C. Meilaender, *Body, Soul, and Bioethics* (Notre Dame:
University of Notre Dame Press, 1996).

—— 2. Foundational Methods ——

1. Stanley G. Clarke and Evan Simpson, in *Anti-Theory in Ethics and
Moral Conservatism* (Albany: SUNY Press, 1989), pp. 3–5. Also, Annette Baier
argues that in modern moral philosophy normative theories are a system of
moral principles in which the less general are derived from the more general.
See "Doing without Moral Theory?" same volume, p. 33.

2. Because this is such an important category in bioethics, chapter 6 of
this book is devoted to the exploration of agreement and consensus.

3. Robert M. Veatch, *A Theory of Medical Ethics* (New York: Basic Books,
1981), pp. 120–26.

4. For example, see Alasdair MacIntyre, *Whose Justice? Which Rationality?* (Notre Dame: University of Notre Dame Press, 1988).

5. For an extensive discussion of the pluralism based on the major moral appeals of consequences, rights, respect for persons, virtues, and justice see Baruch Brody, *Life and Death Decision Making* (New York: Oxford University Press, 1988), pp. 11–55. See also B. Williams, *Ethics and the Limits of Philosophy* (Cambridge, Mass.: Harvard University Press, 1985), pp. 180–83.

6. There are, of course, different meanings and uses of the term "secular." For a careful exploration of the different senses of the term see H. T. Engelhardt, Jr., *Bioethics and Secular Humanism: The Search for a Common Morality* (London: SCM Press, 1991), pp. 22–31. Throughout this book I will use the term to mean a morally neutral framework in which members of different cultures, beliefs, and moral faiths can interact.

7. In this I follow the lead of Saint Augustine, who wrote of our knowledge of truth saying: "'And ye shall know the truth.' Why so? . . . They believed, not because they knew, but that they might come to know. For we believe in order that we may know, we do not know in order that we may come to believe." In *Ionnis evangeliam,* Tractate, 40.9.

8. Some may be struck by my omission of the work of Bernard Gert, Charles Culver, and K. Danner Clouser (see *Bioethics: A Return to Fundamentals* [New York: Oxford University Press, 1997]). The work seems to be clearly crafted as a foundational account. Yet, as a method, it is very similar to the Beauchamp and Childress principlism method in that they assume that there is a "common morality" that is best articulated by a set of rules.

9. See, for example, Sidgwick, *The Methods of Ethics* (Indianapolis: Hackett Publishing Company, 1981) or Alasdair MacIntyre, *After Virtue* (Notre Dame: University of Notre Dame Press, 1981), chapter 6.

10. Peter Singer, *Practical Ethics* (New York: Cambridge University Press, 1994).

11. Singer, pp. 1–8.

12. Singer, pp. 10–11.

13. Singer, p. 11.

14. Singer, p. 12–13.

15. Singer, p. 57.

16. Singer, p. 14.

17. Singer, chapter 2.

18. Singer, pp. 16–26; see especially p. 21.

19. Singer, p. 21.

20. Singer, p. 11.

21. Alan Donagan, *A Theory of Morality* (Chicago: University of Chicago Press, 1977).

22. Donagan, p. 7.

23. Donagan, p. 7.

24. *Babylonian Talmud,* Shabbat, p. 31a.

25. Donagan, p. 52.

26. Donagan, p. 54.

27. Donagan, p. 61.

28. Immanuel Kant, *Grundlegung,* trans. L. W. Beck, pp. 66–67.

29. Donagan, p. 68.

30. See John Chrysostom, "On the Priesthood," Book I, no. 8, in *Nicene and Post-Nicene Fathers,* ed. Philip Schaff, vol. 9 (Grand Rapids, Mich.: Eerdmans, 1956), pp. 37–38. See also Boniface Ramnsey, "Two Traditions on Lying and Deception in the Ancient Church," pp. 504–33.

31. See, for example, Jean-François Lyotard, *The Postmodern Condition,* trans. G. Bennington and B. Mussumi (Manchester: Manchester University Press, 1984); see also H. T. Engelhardt, *Bioethics and Secular Humanism.* One finds a similar account in MacIntyre, *After Virtue.*

32. See MacIntyre, *After Virtue.*

33. See Kevin Wm. Wildes, S.J., "The Ecumenical and Non-Ecumenical Dialectic of Christian Bioethics," *Christian Bioethics* 1 (1995): 121–27.

34. See chapter 4 for more extensive discussion.

35. As another example, the first principle of the natural law according to Thomas Aquinas is to "Do good and avoid evil," *Summa Theologica,* I-II, q. 94, a. 2.

36. See Otfried Hoffe, *Political Justice* (Oxford: Blackwell, 1995).

37. One finds, for example, in Gaius, an appeal to the "law that natural reason establishes among all mankind and which is followed by all people alike, and is called *ius gentium* [law of nations or law of the world] as being the law observed by all mankind. ("Quod uero naturalis ratio inter omnes homines constituit, id apud, omnes populos peraeque custodituruocaturque ius gentium, quasi quo jure omnes gestes utuntur.") *Institutes of Gaius,* trans. Francis De Zulueta, vol. 1 (London: Oxford University Press, 1976), p. 3.

38. Aquinas, *Summa Theologica,* I-II, q. 94, a. 2.

39. J. Finnis, J. Boyle, and G. Grisez, *Nuclear Deterrence, Morality and Realism* (Oxford: Clarendon Press, 1987). Grisez and Boyle wrote an earlier book addressing the euthanasia debate from a natural law tradition. (See G. Grisez and J. Boyle, *Life and Death with Liberty and Justice* [Notre Dame: University of Notre Dame Press, 1979].) I use the more recent book because I think it gives a clearer presentation of the theory and because of its influence in bioethics. See W. May et al., "Feeding and Hydrating the Permanently Unconscious and other vulnerable persons," in *Issues in Law and Medicine* 3 [1987]: 204–17. J. T. McHugh, "Artificially assisted nutrition and hydration," in *Origins* 19 [1989]: 213–16. See also Robert P. George, *Making Men Moral: Civil*

Liberties and Public Morality (Oxford: Clarendon Press, 1995), and John Finnis, *Natural Law and Natural Rights* (Oxford: Clarendon Press, 1980).

40. Finnis, Boyle, Grisez, p. 276. There is, of course, an important discussion about whether or not this interpretation of the natural law tradition by Finnis, Boyle, and Grisez is not a radical re-interpretation of the tradition. Many would argue that the natural law tradition present in Thomas is a tradition that understands the "principles" of morality within a hierarchical teleology of human ends and goods. These interpreters would criticize the "new" interpretation as being more Kantian than Thomistic in its view of moral principles.

41. Finnis, Boyle, Grisez, p. 276.

42. Finnis, Boyle, Grisez, p. 283.

43. MacIntyre, *Whose Justice? Which Rationality?*, p. 396.

44. Finnis, Boyle, Grisez, p. 287.

45. Finnis, Boyle, Grisez, p. 287.

46. Finnis, Boyle, Grisez, p. 288.

47. A central element to the tradition of Roman Catholic moral theology has been the assumption that natural reason, unaided by grace, could "read" nature and know the moral law. Edwin Healy, for example, argues that there is a difference between what is ethically good and evil and that there is " . . . a reason for the difference, and the reason ought to be discoverable" (Healy, *Medical Ethics,* p. 1). The tradition has, by-and-large understood nature in an Aristotelian view as complete and understandable in itself. This view, however, depends on one's understanding of nature and grace [faith]. That is, if nature is understood as complete and grace is a super additum to nature then one will think that natural reason can know the natural law. If, however, as DeLubac and others have argued, grace and nature are mixed, one will not expect that natural reason will be able to know the moral law. Rather one might think, with Paul, that the law is written by the Spirit and is more a matter of discernment than knowledge.

48. " . . . quemlibet matrimonii usum, in quo exercendo, actus, de industria hominum, naturali sua vitae procreandae vi destituatur, Dei et naturae legem infringere . . . ," in *Enchiridion Symbolorum,* ed. H. Denzinger and A. Schonmetzer (Fribourg: Herder, 1963), no. 3717.

49. See, for example, Joseph Fuchs, S.J., "The Absoluteness of Moral Terms," in C. Curran and R. McCormick, *Readings in Moral Theology, No. 1* (New York: Paulist Press, 1979), pp. 94–137.

50. See Hoffe, p. 50 ff.

51. See, for example, John Locke, *The Second Treatise of Government* (1690, reprint, Indianapolis: Hackett, 1980); Thomas Hobbes, *Leviathan;* E. Barkes, ed., *Social Contract: Essays by Locke, Hume, and Rousseau* (Oxford: Oxford University Press, 1948).

52. See Veatch, *A Theory of Medical Ethics,* for his complete theory. Also see Veatch, "Resolving Conflicts among Principles: Ranking, Balancing, and Specifying," *Kennedy Institute of Ethics Journal* 5: 199–218.

53. Veatch, *A Theory,* p. 137.

54. Veatch, pp. 137–38.

55. Veatch, Part 3.

56. Veatch, pp. 127–34.

57. Veatch, pp. 134–37.

58. See Veatch, chapters 6 and 11.

59. Veatch, p. 299.

60. Veatch, p. 304.

61. N. Daniels, *Just Health Care* (Cambridge: Cambridge University Press, 1985).

62. John Rawls, *A Theory of Justice* (Cambridge, Mass.: Harvard University Press, 1971).

63. See chapter 3. Also, T. Beauchamp and J. Childress, *The Principles of Biomedical Ethics,* 3d ed. (New York: Oxford University Press, 1989).

64. For a discussion of this topic see Christopher Boorse, "Health as a Theoretical Concept," *Philosophy of Science* 44 (1977): 542–71; Joseph Margolis, "The Concept of Disease," *The Journal of Medicine and Philosophy* 1 (1976): 238–55; H. T. Engelhardt, Jr., "Ideology and Etiology," *Journal of Medicine and Philosophy* 1 (1976): 256–58.

65. Veatch, p. 138.

66. Pellegrino and Thomasma, *A Philosophical Basis of Medical Practice.*

67. Pellegrino and Thomasma, *Philosophical Basis,* pp. 80–81.

68. Pellegrino and Thomasma, p. 219.

69. Pellegrino and Thomasma, p. 148; Aristotle, *Metaphysics* 1140b5–6.

70. Pellegrino and Thomasma, pp. 177–80.

71. Pellegrino and Thomasma, pp. 192–220.

72. Edmund D. Pellegrino and David C. Thomasma, *For the Patient's Good: The Restoration of Beneficence in Health Care* (New York: Oxford University Press, 1988).

73. Pellegrino and Thomasma, *For the Patient's Good,* pp. 111–24.

74. Pellegrino and Thomasma, *A Theory,* pp. 50–56.

75. Timothy E. Quill, "Death and Dignity: A Case of Individualized Decision Making," *New England Journal of Medicine* 324, no. 10 (March 7, 1991): 691–94.

76. See MacIntyre, *After Virtue.*

77. This point is often raised in feminist literature that critiques the models of rationality deployed in bioethics. (See Alia L. Carse, "The 'Voice of Care': Implications for Bioethical Education," *Journal of Medicine and Philosophy* 16 (1991): 5–28.) Often, however, the critics of traditional models of

reason fail to grasp the implications of their views in that *all* accounts of reason suffer the same difficulty in a secular, pluralistic setting.

78. G. F. W. Hegel, *Philosophy of Right,* trans. T. M. Knox (London: Oxford University Press, 1967), third part.

79. See H. T. Engelhardt, Jr., *The Foundations of Bioethics,* 2d ed. (New York: Oxford University Press, 1996).

80. See Baruch A. Brody, *Moral Theory and Moral Judgments in Medical Ethics* (Dordrecht: Kluwer, 1988).

81. A comprehensive, theological expression of the natural law, as the framework of moral theology, was not found in the Christian tradition until the thirteenth century. See, for example, Aquinas, *Summa Theologica,* I-II, q. 90–96.

82. For example, one finds in Healy's medical ethics the first postulate for ethical principles to be the acceptance of "the existence of a Creator as a *truth* proved in philosophy," Healy, *Medical Ethics,* pp. 2–3.

83. See Tom L. Beauchamp and James F. Childress, *The Principles of Biomedical Ethics,* 4th ed. (New York: Oxford University Press, 1994).

84. See, for example, Albert Jonsen and Stephen Toulmin, *The Abuse of Casuistry: A History of Moral Reasoning* (Berkeley: University of California Press, 1988).

—— 3. ECUMENISM IN BIOETHICS ——

1. Unless otherwise noted, Beauchamp and Childress, *The Principles of Biomedical Ethics,* 4th ed. is quoted in this chapter.

2. It is important to note that the Beauchamp and Childress project is very similar to that of Jonsen and Toulmin (*The Abuse of Casuistry*). Both recognize the foundational difficulties. Both think these difficulties can be sidestepped (in different ways) so as to facilitate the resolution of moral controversies in bioethics. In the analysis and argument that follow I will take the position that both projects fail because of the same issues that plague foundationalism.

3. Beauchamp and Childress wrote in a footnote of the third edition that while there are differences one also finds a great deal of common history (p. 24, n. 20). In the fourth edition they have explicated these assumptions.

4. See, Engelhardt, Jr., *The Foundations of Bioethics,* 2d ed. (New York: Oxford University Press, 1996).

5. Beauchamp and Childress, p. 5.

6. Beauchamp and Childress, p. 13.

7. Beauchamp and Childress, p. 15.

8. Beauchamp and Childress, p. 14.

9. Beauchamp and Childress, p. 17.
10. Beauchamp and Childress, p. 17.
11. Beauchamp and Childress, p. 18.
12. Beauchamp and Childress, p. 19.
13. Beauchamp and Childress, p. 20.
14. Beauchamp and Childress, p. 24.
15. Beauchamp and Childress, p. 25.
16. Beauchamp and Childress, p. 37.
17. Beauchamp and Childress, p. 38.
18. Beauchamp and Childress, chapter 2.
19. Beauchamp and Childress, p. 38.
20. Beauchamp and Childress, p. 102.
21. Beauchamp and Childress, pp. 104–6.
22. Beauchamp and Childress, p. 111.
23. Beauchamp and Childress, pp. 32–37.
24. Beauchamp and Childress, p. 29.
25. Beauchamp and Childress, p. 120.
26. Beauchamp and Childress, p. 121.
27. Beauchamp and Childress, pp. 122–23.
28. Beauchamp and Childress, p. 123.
29. Beauchamp and Childress, p. 123.
30. Beauchamp and Childress, pp. 190–93.
31. Beauchamp and Childress, pp. 190–93.
32. Beauchamp and Childress, p. 192.
33. Beauchamp and Childress, p. 193.
34. Beauchamp and Childress, pp. 194–96.
35. Beauchamp and Childress, p. 260.
36. Beauchamp and Childress, p. 260.
37. Beauchamp and Childress, p. 259.
38. Beauchamp and Childress, pp. 261–71.
39. Beauchamp and Childress, p. 268.
40. Beauchamp and Childress, pp. 266–71.
41. Beauchamp and Childress, pp. 330–31.
42. Beauchamp and Childress, p. 328.
43. T. C. Sandars, ed., *The Institutes of Justinian* (Westport, Conn.: Greenwood Press, 1922), Book I, 1.
44. Beauchamp and Childress, p. 330.
45. Beauchamp and Childress, p. 330.
46. Beauchamp and Childress, p. 395.
47. Beauchamp and Childress, p. 396.
48. Beauchamp and Childress, p. 142 ff.

49. *Mohr v. Williams,* 95 Minn. 261, 104 N.W. 12 (1905), at 15; see Beauchamp and Childress, p. 143.

50. Beauchamp and Childress, pp. 144–46.

51. Beauchamp and Childress, p. 125.

52. See Tom L. Beauchamp, "Physician-Assisted Suicide: A Response to Edmund Pellegrino" in *Choosing Life: A Dialogue on* Evangelium Vitae, ed. K. Wildes, S.J., and A. Mitchell (Washington, D.C.: Georgetown University Press, 1997), pp. 254–58.

53. It should be noted that while Childress is a deontologist he does not endorse a Kantian deontology.

54. Beauchamp and Childress, p. 125.

55. H. T. Engelhardt, Jr., *The Foundations of Bioethics;* R. Nozick, *Anarchy, State, and Utopia* (New York: Basic Books, 1974).

56. See Joel Feinberg, *Harm to Others: The Moral Limits of the Criminal Law* (New York: Oxford University Press, 1984), chapter 1, pp. 31–64.

57. Beauchamp and Childress, pp. 259–60.

58. Beauchamp and Childress, p. 194.

59. Beauchamp and Childress, pp. 386–87.

60. Beauchamp and Childress, pp. 38–40.

61. Beauchamp and Childress, p. 218.

62. J. F. Childress and C. C. Campbell, "'Who Is a Doctor to Decide Whether a Person Lives or Dies?' Reflections on Dax's Case," in *Dax's Case,* ed. L. D. Kliever (Dallas: Southern Methodist University Press, 1989), pp. 23–41.

63. Beauchamp and Childress, p. 104 ff.

64. Beauchamp and Childress, pp. 104–5.

65. Beauchamp and Childress, pp. 273–75.

66. Beauchamp and Childress, p. 212.

67. See the discussion of casuistry and considered judgments in chapters 1 and 2.

68. If one looks at different "texts" in the field of bioethics one finds different principles cited and used. Ronald Munson, for example, cites principles of autonomy, beneficence, distributive justice, nonmaleficence, and utility (see Ronald Munson, *Intervention and Reflection: Basic Issues in Medical Ethics,* 4th ed. [Belmont, Calif.: Wadsworth, 1992]).

69. See Beauchamp and Childress, pp. 190–91, and their discussion of nonmaleficence.

70. Beauchamp and Childress, p. 24, n. 20.

71. Beauchamp and Childress, p. 5.

72. Aristotle, *Metaphysics* 1013b20.

73. Healy, *Medical Ethics,* p. 5.

74. Aquinas, *Summa Theologica,* I-II, 94, 2.

75. Aquinas, I-II, 94, 2.

76. Aquinas, I-II, 94, 2.

77. Aquinas, I-II, 94, 4–5.

78. James F. Childress, "The Identification of Ethical Principles," *Journal of Religious Ethics* 5 / 1 (1977): 39–68; R. M. Hare, "Principles," in *Essays in Ethical Theory* (Oxford: Clarendon Press, 1989).

79. Childress, p. 41.

80. Childress, p. 58.

81. J. Rawls, "Two Concepts of Rules," *Philosophical Review* 64 (1955): 3–32.

82. L. Wittgenstein, *Philosophical Investigations,* trans. G. E. M. Anscombe (New York: Macmillan, 1958), sections 191, 201, 202, 206, 217–19, 223.

83. J. Rawls, "Justice as Fairness: Political not Metaphysical," *Philosophy and Public Affairs* 14 / 3 (1985): 223–51.

—— 4. After Paradigms ——

1. Two of the most prominent examples of casuistry in bioethics are found in Jonsen and Toulmin, *The Abuse of Casuistry* and Brody, *Life and Death Decision Making.*

2. The literature about the "Trolley Paradox," which will be addressed later in the chapter, provides such an example. See Philippa Foot, "The Problem of Abortion and the Doctrine of the Double Effect," *Oxford Review* (1967); Judith Jarvis Thompson, "The Trolley Problem," *Yale Law Journal* 94 (May 1985): 1395–1415; F. M. Kamm, "Harming Some to Save Others," *Philosophical Studies* 57 (1989): 227–60.

3. The National Commission for the Protection of Human Subjects of Biomedical and Behavioral Research, *The Belmont Report: Ethical Principles and Guidelines for the Protection of Human Subjects of Research* (Washington, D.C.: U.S. Government Printing Office, 1978), DHEW Publication No. (OS) 78-0012.

4. Jonsen and Toulmin, p. 27.

5. Jonsen and Toulmin, p. 26.

6. Jonsen and Toulmin, p. 34.

7. Jonsen and Toulmin, p. 37. See also Kathryn Montgomery Hunter, "A Science of Individuals: Medicine and Casuistry," *Journal of Medicine and Philosophy* 14 (1989): 193–221.

8. Jonsen and Toulmin, pp. 44–45.

9. Jonsen and Toulmin, p. 4.

10. Jonsen and Toulmin, p. 6.

11. See Stephen Toulmin, "The Tyranny of Principles," *Hastings Center Report* 11 (December 1981): 31–39.

12. Jonsen and Toulmin, p. 10.

13. Jonsen and Toulmin, p. 13.

14. Jonsen and Toulmin, p. 66.

15. Jonsen and Toulmin, p. 72.

16. Jonsen and Toulmin, pp. 76–77.

17. Jonsen and Toulmin, p. 79; Cicero, *De Officiis,* I, 59.

18. Jonsen and Toulmin, p. 83.

19. Jonsen and Toulmin, p. 88.

20. Jonsen and Toulmin, pp. 92–96.

21. "Bigotan Penitential" in *The Irish Penitentials,* ed. L. Bieler (Dublin: Trinity Press, 1963), p. 199.

22. Jonsen and Toulmin, p. 100.

23. *Unam Sanctum,* November 13, 1302; see *Enchiridion Symbolorum,* ed. H. Denzinger and A. Schonmetzer (New York: Herder, 1963), nos. 870–75.

24. Jonsen and Toulmin, p. 123.

25. Jonsen and Toulmin, p. 127; Aquinas, *Summa Theologica,* I-II, q. 91, a. 1–2.

26. Jonsen and Toulmin, p. 135.

27. For example, the whole dispute over probabilism was an argument about the authority of the opinions of moral theologians when a matter was unsettled or unclear.

28. Jonsen and Toulmin, p. 307.

29. Jonsen and Toulmin, pp. 306–7.

30. Jonsen and Toulmin, p. 309.

31. Jonsen and Toulmin, pp. 311–12.

32. Jonsen and Toulmin, pp. 312–13.

33. Jonsen and Toulmin, pp. 314–16.

34. Jonsen and Toulmin, pp. 329–31.

35. Jonsen and Toulmin, p. 306.

36. M. Masterman, "The Nature of a Paradigm," in *Criticism and the Growth of Knowledge,* ed. I. Lakatos and A. Musgrave (London: Cambridge University Press, 1970), pp. 59–89.

37. R. Grandy, "Incommensurability: Kinds and Causes," *Philosophica* 32 (1983): 8.

38. Grandy, pp. 8–9.

39. *Summa Theologica,* II-II, q. 64, a. 7. In his discussion of self-defense Aquinas speaks of actions being identified in species, according to the intention of the agent. In I-II, q. 7, a. 1, Thomas speaks of circumstances of an act as the accidents of an act and being outside its substance.

40. Some argue that casuistry arose, in part, as a response to questions about the metaphysics of moral acts and that many of the casuists of sixteenth and seventeenth century were nominalists (see Diego Gracia, "The

Old and the New in the Doctrine of the Ordinary and the Extraordinary Means," in *Critical Choices and Critical Care: Catholic Perspectives on Allocating Resources in Intensive Care Medicine,* ed. K. Wildes, S.J. [Dordrecht: Kluwer, 1995], pp. 119–125). This account of the period of high casuistry serves to highlight the role of juridical authority in settling moral disputes. How we account for moral authority in contemporary, morally pluralistic societies will be instructive as to the possibilities for casuistry in this context.

41. Grandy, pp. 10–11.

42. Augustine writes: "I do not agree with the opinion that one may kill a man if one may be killed by him; unless one be a soldier, exercising a public office, so that one dies it is not for oneself but for others, having the power to do so. . . . " Letter to Publicola, no. 48.

43. Laurence H. Tribe, *Abortion: The Clash of Absolutes* (New York: W. W. Norton, 1990).

44. Richard E. Flathman, "Power, Authority, and Rights in the Practice of Medicine," in *Responsibility in Health Care,* ed. George Agich (Dordrecht: Reidel, 1982), 108–10.

45. J. A. McHugh, O.P., *The Casuist: A Collection of Cases in Moral and Pastoral Theology,* vol. 5 (New York: Joseph Wagner, 1917), pp. 85–87.

46. Thomas N. Tentler, "The *Summa* for Confessors and an Instrument of Social Control," in *The Pursuit of Holiness in Late Medieval and Renaissance Religion,* ed. C. Trinkaus and H. Oberman (Leiden: E. J. Brill, 1974), pp. 103–37.

47. J. Neuner and J. Dupuis, eds., *The Christian Faith* (New York: Alba House, 1982), no. 1608; Denzinger and Schonmetzer, no. 812.

48. Pierre J. Payer, *Sex and the Penitentials* (Toronto: University of Toronto Press, 1984).

49. Darrell W. Amundsen, "Casuistry and Professional Obligations: The Regulation of Physicians by the Court of Conscience in the Late Middle Ages," *Transactions and Studies of the College of Physicians of Philadelphia* 3, no. 2 (March 1981): 22–29.

50. See Tentler.

51. J. Mahoney, *The Making of Moral Theology: A Study of the Roman Catholic Tradition* (New York: Oxford University Press, 1987), pp. 17–22.

52. One can look at the responses in recent years to Pope John Paul II's letter *Ex Corde Ecclesiae* as an example of the continuing tensions between universities and church authorities.

53. James A. Gallagher, *Time Past, Time Future: An Historical Study of Catholic Moral Theology* (New York: Paulist Press, 1990), pp. 5–28.

54. Denzinger and Schonmetzer, no. 1688.

55. Mahoney, p. 19.

56. Denzinger and Schonmetzer, no. 813, emphasis added.

57. *Nicomachean Ethics* 6.13, 1144b17–1145a6; 10.8, 1178a16–19.

58. Richard Sorabji, "Aristotle on the Role of Intellect in Virtue," in *Essays on Aristotle's Ethics,* ed. Amelie Rorty (Berkeley: University of California Press, 1980), p. 205.

59. Sarah Broadie, *Ethics with Aristotle* (New York: Oxford University Press, 1991), p. 258.

60. *NE* 6.12, 1144a22–b1.

61. *NE* 6.1. It is such knowledge that Jonsen and Toulmin's account ignores.

62. M. T. Cicero, *De Officiis,* trans. Walter Muller (Cambridge, 1913), I, 43.

63. *Summa Theologica,* II–II, 47, 5; I–II q. 47, aa. 1–16.

64. *Summa Theologica,* II–II, q. 47, a. 6.

65. Saint Ignatius of Loyola, *The Constitutions of the Society of Jesus,* trans. George E. Ganss, S.J. (St. Louis: The Institute of Jesuit Sources, 1970), no. 582, pp. 260–61, n. 2.

66. Saint Ignatius of Loyola, *The Spiritual Exercises,* no. 176, 328.

67. "They show that what the law requires is written on their hearts . . . ," Romans 2:15.

68. Ignatius, *Constitutions,* no. 134; *Summa Theologica,* I–II, q. 90; 92, a. 1.

69. See, for example, Healy, *Medical Ethics,* pp. 98–101; Alphonsus de Ligorio, *Homo Apostolicus* (Torino: Tipografia Pontificia, 1890), pp. 174–77. There has been an ongoing discussion about the principle of double effect and its relationship to an absolutist morality. The discussion highlights the idea that any principle needs to be understood in the web of moral commitments. See, for example, J. Boyle, "Who Is Entitled to Double Effect," *Journal of Medicine and Philosophy* 16 (1991): 475–94; Mark Aulisio, "One Person's Modus Ponens: Boyle, Absolutist Catholicism, and the Doctrine of Double Effect," *Christian Bioethics* 3 (1997): 142–57.

70. See Joseph Boyle, "Who Is Entitled to Double Effect," pp. 475–94.

71. Jonsen and Toulmin, p. 307.

72. Bartolomeo Medina, *Expositio in Summae Theologiae Partem,* I–II, q. 19.

73. H. Davis, *Moral and Pastoral Theology,* vol. 1 (New York: Sheed and Ward, 1936), pp. 82–113.

74. G. Vasquez, *Commentaria,* II–II, LXII, iv.

75. Jonsen and Toulmin, p. 170.

76. I would argue that the turn to the individual person as the source of authority (in morality, politics, aesthetics, etc.) is one of the marks of the modern age. The emergence of postmodernity, I think, comes about when there is no longer a common, dominant culture that brings individuals together. One might argue that postmodernism is the logical outcome of modernity.

77. Saint Ambrose, *De Tobia,* XV, 15; First Council of Nicea, Canon 15; Gratian's *Decretum* II, c. 14, 4.

78. Saint Jerome, *Commentarium on Ezechiel,* VI:18.

79. *Summa Theologica,* II-II, 78, a. 1.

80. Jonsen and Toulmin, p. 187.

81. See MacIntyre, *Whose Justice? Which Rationality?,* pp. 349–69.

82. T. Beauchamp, pp. 30–31.

83. E. Leites, "Casuistry and Character," in *Conscience and Casuistry in Early Modern Europe,* ed. Edmund Leites (Cambridge: Cambridge University Press, 1988), pp. 185–213.

84. Jonsen and Toulmin, p. 327.

85. Jonsen and Toulmin, p. 327.

86. Henrik R. Wulff, *Rational Diagnosis and Treatment: An Introduction to Clinical Decision-Making* (Oxford: Blackwell Scientific Publications, 1981), p. 69.

87. Jonsen and Toulmin, pp. 306–7.

88. Engelhardt, *The Foundations of Bioethics.*

89. See Kamm, p. 227; Thompson, p. 103.

90. Kamm, p. 227, emphasis added.

91. MacIntyre, appendix 1 to *The Belmont Report: Ethical Principles and Guidelines for the Protection of Human Subjects of Research,* pp. 10–14.

92. J. J. Thompson, "A Defense of Abortion," *Philosophy and Public Affairs,* vol. 1, no. 1 (1971): 47–66.

93. See, for example, Baruch Brody, *Abortion and the Sanctity of Life: A Philosophical View* (Cambridge, Mass.: MIT Press, 1975), p. 26 ff.

94. Elizabeth Wolgast, "Moral Pluralism," *The Journal of Social Philosophy* 21, no. 2 and 3 (1990): 108–16.

95. While I do think that objective moral standards do exist, I also think there are profound epistemological difficulties in knowing what these standards are.

96. John Arras, "Review—Common Law Morality," *Hastings Center Report* 20 (July/August 1990): 35–37.

97. William H. Gass, "The Case of the Obliging Stranger," *The Philosophical Review* 66 (1957): 202.

98. See Gass, pp. 193–204.

—— 5. COMMUNITARIAN BIOETHICS ——

1. See, for example, Ezekiel Emanuel, *The Ends of Human Life: Medical Ethics in a Liberal Polity* (Cambridge: Harvard University Press, 1991). See also E. Emanuel, "A Communitarian Health-Care Package," in *Responsive Community,* Summer 1993, 49–56.

2. See, for example, MacIntyre, *After Virtue.*

3. See Meilaender *Body, Soul, and Bioethics,* pp. 37–60.

4. While I am sympathetic to Emanuel's project I think he underestimates the importance and degree of moral pluralism in a secular society such as the United States in his effort to develop a content-full bioethics for the nation. Along this line of thought, see, for example, Daniel Callahan, "Communitarian Bioethics: A Pious Hope?" *Responsive Community* 6/4 (Fall 1996): 26–33; Arthur Caplan, "Communitarian Bioethics: Reasons for Pessimism," *Responsive Community* 7/1 (Winter 1996–97): 93–95.

5. Engelhardt, *The Foundations of Bioethics,* p. 74.

6. Engelhardt, *The Foundations of Bioethics,* p. 77.

7. The penalty is as follows: "A person who procures a successful abortion incurs an automatic *(latae sententiae)* excommunication." (A *latae sententiae* penalty is a penalty inflicted by the law itself upon commission of the offense. It is distinguished from a *ferendae sententiae* penalty, which is imposed by the action of a judge or superior.) See *Code of Canon Law* (Washington, D.C.: Canon Law Society of America, 1983), c. no. 1398.

8. See George Marsden, *The Soul of the American University: From Protestant Establishment to Established Non Belief* (New York: Oxford University Press, 1994). Marsden argues that Protestant universalism became the very ideology that undermined strong Christian beliefs in Protestant universities.

9. See Lisa S. Cahill, "Theology and Bioethics: Should Religious Traditions Have a Public Voice?" *The Journal of Medicine and Philosophy* 17: 263–72.

10. This may help to explain why Roman Catholics often practice abortion at the same rate as the general population.

11. S. Hauerwas, *The Peaceable Kingdom: A Primer in Christian Ethics* (Notre Dame: University of Notre Dame Press, 1984), p. 85.

12. In Roman Catholic thought one is struck as to how its deep commitment to rationalism, especially in its moral theology, is also tied to its authoritarianism. When reason fails to persuade, the tradition must resort to the assertion of authority. But the use of authority is *ad hoc* in that it comes from outside the moral tradition since the moral tradition appeals to the authority of reason.

13. See, for example, Richard John Neuhaus, *The Naked Public Square: Religion and Democracy in America* (New York: Eerdmans, 1986); Stephen L. Carter, *The Culture of Disbelief: How American Law and Politics Trivialize Religious Devotion* (New York: Basic Books, 1993).

14. See, for example, Abraham Steinberg, "Jewish Medical Ethics" in *Theological Developments in Bioethics: 1988–1990,* ed. B. A. Brody, B. A. Lustig, H. T. Engelhardt, Jr., L. B. McCullough (Dordrecht: Kluwer, 1991), pp. 179–99.

15. Engelhardt, *Bioethics and Secular Humanism,* p. 3 ff.

16. Engelhardt, *The Foundations of Bioethics,* p. 80.

17. K. Wildes, S.J., "Engelhardt's Communitarian Ethics: The Hidden Assumptions," in *Reading Engelhardt,* ed. Brendan Minogue et al. (Dordrecht: Kluwer, 1997).

18. In his recent book, Erich Loewy has developed his own understanding of moral acquaintances. Loewy's position is different from my own in certain key respects. He assumes that Engelhardt is a libertarian who views the individual as isolated. As I have argued elsewhere, I think such a view is mistaken (see note 16). Loewy, like many others in bioethics, begins by assuming an area of moral agreement that he calls "connectedness." I think it is more helpful to move in the opposite direction and begin with differences in hopes of finding what it is we hold in common. See Loewy, *Moral Strangers, Moral Acquaintance, and Moral Friends.*

19. Mary Warnock, *Report on the Committee of Inquiry into Human Fertilisation and Embryology* (London: Her Majesty's Stationery Office, 1984), p. xi.

—— 6. THE GENEALOGY OF AGREEMENT AND CONSENSUS ——

1. Throughout this chapter the terms "consensus" and "agreement" will be used interchangeably as they often are in bioethics. Both words have Latin roots in the expression of "being of one heart." It is clear that agreement is a necessary condition for consensus.

2. Jonathan D. Moreno, *Deciding Together* (New York: Oxford University Press, 1995).

3. See Nicholas Rescher, *Pluralism: Against the Demand for Consensus* (New York: Oxford University Press, 1993).

4. Peter Singer, *Practical Ethics* (Cambridge: Cambridge University Press, 1993), pp. 135–74.

5. L. Wittgenstein, *Philosophical Investigations,* no. 66 and 67.

6. Rescher, p. 15.

7. Rescher, p. 93.

8. Rescher, p. 105.

9. See Kamm, "Harming Some to Save Others."

10. See chapter 3.

11. The debate on the liceity of craniotomy to save the life of the mother went on among theologians and casuists in the nineteenth century. In 1872, the Sacred Penitentiary, when asked if a craniotomy could be preformed, answered, "consult approved authors, old and new, and act prudently." Old and approved authors like Sanchez could have been interpreted as authorizing craniotomies. By 1889, however, the Penitentiary wrote that it was not safe to teach that a craniotomy to save the mother's life was lawful (Denzinger, no. 1889). This historical example is instructive as it illustrates

the interplay between the content of a particular tradition and the authority structure of the tradition.

12. *Summa Theologica,* II-II, q. 47.

13. For a discussion of the range of possible responses to abortion see V. Genovesi, "Challenging the Legal Status of Abortion: A Matter of Moral Obligation?" *America* 153 (1985): 417–22.

14. Kelly, *Medico-Moral Problems,* pp. 139–43.

15. Pius XII, "Tissue Transplantation, May 14, 1956," in *The Human Body: Papal Teaching,* ed. Monks of Solesmes (Boston: Daughters of St. Paul, 1960).

16. See Jonathan Moreno, "Consensus, Contracts, and Committees," *The Journal of Medicine and Philosophy* 16 (1991): 393–408; J. Moreno, "Consensus by Committee: Philosophical and Social Aspects of Ethics Committees," in *The Concept of Moral Consensus: The Case of Technological Interventions into Human Reproduction,* ed. K. Bayertz (Dordrecht: Kluwer, 1994), pp. 145–62.

17. See James F. Childress, "Consensus in Ethics and Public Policy: The Deliberations of the U.S. Human Fetal Tissue Transplantation Research Panel," in Bayertz, *The Concept of Moral Consensus,* pp. 163–87.

18. Childress, *Consensus in Ethics,* p. 170.

19. Childress, *Consensus in Ethics,* p. 165.

20. Childress, *Consensus in Ethics,* p. 169.

21. Childress, *Consensus in Ethics,* p. 165 ff.

22. Norman Daniels, "The Articulation of Values and Principles Involved in Health Care Reform," *The Journal of Medicine and Philosophy* 19 (1994): 425–34.

—— 7. MORAL ACQUAINTANCES ——

1. See Michael Sandel, *Democracy's Discontent: America in Search of a Public Philosophy* (Cambridge: The Belknap Press, 1996).

2. I am deeply indebted to my conversations with Professor Terry Pinkard on this point.

3. See Sandel.

4. See Kevin Wm. Wildes, S.J., "Conscience, Referral, and Physician Assisted Suicide," *Journal of Medicine and Philosophy* 18 (1993): 323–28.

5. J. C. Murray, *We Hold These Truths: Catholic Reflections on the American Proposition* (New York: Sheed and Ward, 1960), p. 33.

6. *Cruzan v. Harmon,* 760 S.W.2d 408 (Mo. Banc 1988).

7. One sees this in the history of the United States. Though conceived as a secular state it has traditionally conceived itself as a Christian nation. The Supreme Court wrote that "we are a Christian people," *Church of the Holy*

Trinity v. United States, 143 US 457 (1892). In this case the Court held that the Christian religion was part of the common law of Pennsylvania.

8. W. Blackstone, *Blackstone's Commentaries,* ed. St. George Tucker, vol. 1 (New York: August M. Kelley, 1969), p. 40.

9. Blackstone, *Commentaries,* vol. 5, pp. 42–55.

10. See also P. Devlin, *The Enforcement of Morals* (London: Oxford University, 1969); M. Cathleen Kaveny, "The Limits of Ordinary Virtue: The Limits of the Criminal Law in Implementing *Evangelium Vitae,*" in *Choosing Life: A Dialogue on* Evangelium Vitae, ed. K. Wildes, S.J., and A. Mitchell (Washington: Georgetown University Press, 1997), pp. 134–51; also Kevin P. Quinn, S.J., "Whose Virtue? Which Morality? The Limits of Law As a Teacher of Virtue," in *Choosing Life,* ed. Wildes and Mitchell, pp. 152–57.

11. Edmund Pellegrino, "Doctors Must Not Kill," *Journal of Clinical Ethics* 3 (1992): 95–102.

12. T. Quill, C. Cassel, D. Meier, "Care of the Hopelessly Ill: Proposed Clinical Criteria for Physician-Assisted Suicide," *New England Journal of Medicine* 327 (1992): 1382.

13. Wildes, "Conscience, Referral, and Physician Assisted Suicide."

14. The cult of "political correctness" has overlooked an old Roman Catholic maxim in confessional practice: One should love the sinner but hate the sin. Thus one can certainly love the sinner and tolerate their sin but one should hate the sin and denounce it for what it is. Perhaps we would do better to speak of "nonviolent intolerance" to stress the difference between toleration and approval.

15. In the *millet* system the subject class could organize themselves as they wished in accord with religious faith or occupation. The system allows for each group to develop and implement its own rules about marriage, divorce, birth, health, and education.

16. See, for example, George Khushf, "Administrative and Organizational Ethics," *HEC Forum* 9 (1997): 299–309.

17. Donald L. Madison, "Preserving Individualism in the Organizational Society: 'Cooperation' and American Medical Practice, 1900–1920," *Bulletin of the History of Medicine* 70 (1996): 442–83.

18. Ellen Fox, "Predominance of the Curative Model of Medical Care: A Residual Problem," *Journal of the American Medical Association* 278: 761–63.

19. See, for example, J. Collins and J. I. Porras, "Building Your Company's Vision," *Harvard Business Review* (September-October 1996): 65–77.

20. United States Catholic Conference, "Ethical and Religious Directives for Catholic Health Care Services," *Origins* 24:449, 451–62.

21. Traditional Roman Catholic thought had distinguished between three different spheres of justice: commutative, distributive, and legal. In his 1931 encyclical, *Quadragesimo Anno,* Pius XI introduced the term "social jus-

tice." Though its meaning is unclear the term seems to be tied to the real-ization of the common good in a society and challenges a society to examine the way it is organized.

22. See Meilaender, *Body, Soul, and Bioethics.*

23. See Wickler, "Federal Bioethics: Methodology, Expertise and Evalu-ation."

24. T. S. Eliot, "Four Quartets," in *The Complete Poems and Plays* (New York: Harcourt, Brace, and World, 1971), p. 144.

25. See chapter 1, note 3 above. As I have noted elsewhere, Roman Catholics have traditionally assumed that almost all of the particular content of bioethics can in fact be established by reason alone.

INDEX

abortion: Catholic view of, 130, 134–35, 176; Jonsen/Toulmin on, 91–92; Judeo-Christian commonality and, 83; justification in plural society, 134; moral friends/acquaintances on, 18, 19; no common morality for, 35, 139–40; paradigms for, 97; pluralism and, 77; principlism and, 75; Roman vs. Orthodox Catholics, 150–51; rule of law and, 168; Thompson on, 117
absolutism, 92, 93
abstraction, concept of, 80
After Virtue (MacIntyre), 75
agreement (consensus): in casuistry, 92, 118–19; content lost by, 133; contractarian view of, 45; on essence of disputes, 154; foundationalism and, 24–25; between friends/strangers/acquaintances, 18–19, 78, 157–58; meaning and goal of, 142–61; Moreno on, 142–43; necessary to proceduralism, 163; need for, 53–54; pluralism and, 145–47; principlism's view of, 61; role in bioethics, 179; sociology of, 152–57; spectrum and levels of, 147–53; use of terms, 200n1; in Veatch's model, 43
Amish option, 133, 136, 172
Aristotle: concept of justice, 64–65; on first principles, 79; on moral knowledge, 91, 93; on prudence, 105; view of nature, 189n47; on virtues, 47, 50
Augustine, 34–35, 99, 187n7, 196n42

authority: of bioethicist, 180; casuistry's reliance on, 87, 88–89, 92, 97, 98, 100–101; in Christianity, 52, 81; communities' views of, 127, 131–32; function of, 131; individuals as source of, 108, 111, 197n76; juridical authority, 101, 111; limits of state, 168; in probabilism, 106–9; in proceduralism, 16; role in Renaissance casuistry, 102–5, 107, 110–11, 112–14; role in secular casuistry, 114–15, 119; moral sources of, 165–66; Suarez on, 108; of texts, 130–31; tool for understanding, 136–37
autonomy: bioethics reduced to, 15; communitarian view of, 125; meaning of, 67–69; of patients, 19, 48–51; principle of (respect for), 62, 63, 65, 67–69, 72, 73; rise of, 111

balancing of principles: coherence from, 80; in contractarianism, 46; pluralism and, 81; in principlism, 62, 66, 70–73, 76–77; process of, 55, 57; ranking, 79–80
Beauchamp, Tom L.: on common morality, 26, 35, 169; contractarianism and, 45; limitations of theory, 191n2; method of, 53; on National Commission, 90, 142; on PAS, 19; principlism of, 20, 55–85
Belmont Report, 90
beneficence: meaning of, 69–70; paternalism and, 73; of physicians, 48;

205